T0211668

Lecture Notes of the Institute for Computer Sciences, Social Informatics and Telecommunications Engineering   400

More information about this series at http://www.springer.com/series/8197

Youssou Faye · Assane Gueye · Bamba Gueye ·
Dame Diongue · El Hadji Mamadou Nguer ·
Mandicou Ba (Eds.)

Research in
Computer Science
and Its Applications

11th International Conference, CNRIA 2021
Virtual Event, June 17–19, 2021
Proceedings

 Springer

Editors
Youssou Faye ⓘ
Université Assane Seck de
Ziguinchor, Senegal

Assane Gueye ⓘ
Carnegie Mellon University Africa
Kigali, Rwanda

Bamba Gueye ⓘ
Université Cheikh Anta Diop
Dakar, Senegal

Dame Diongue ⓘ
Université Gaston Berger de Saint Louis
Saint Louis, Senegal

El Hadji Mamadou Nguer
Virtual University of Senegal UVS
Dakar, Senegal

Mandicou Ba
Ecole Supérieure Polytechnique
Dakar, Senegal

ISSN 1867-8211   ISSN 1867-822X (electronic)
Lecture Notes of the Institute for Computer Sciences, Social Informatics
and Telecommunications Engineering
ISBN 978-3-030-90555-2   ISBN 978-3-030-90556-9 (eBook)
https://doi.org/10.1007/978-3-030-90556-9

This Springer imprint is published by the registered company Springer Nature Switzerland AG
The registered company address is: Gewerbestrasse 11, 6330 Cham, Switzerland

Preface

We are delighted to introduce the proceedings of the eleventh edition of the Conference on Research in Computer Science and Its Applications (CNRIA 2021). CNRIA is an international scientific conference that highlights new trends and advances in current topics in the field of computing and their applications for development. CNRIA is organized by the Senegalese Association of Researchers in Computer Science (ASCII). The main purpose of the CNRIA 2021 conference was to serve as a forum for researchers from academia, the ICT industry, and development actors to discuss recent advances and cutting-edge topics in the field of computer science and its applications. This conference also served as a forum for researchers and practitioners to exchange ideas, propose solutions, discuss research challenges, and share experience. CNRIA 2021 goals included promoting the popularization of the contributions of the regional scientific community in new cutting-edge research topics such as AI, IoT, big data, security, and SDN.

This 11th edition of CNRIA was held during June 17–19, 2021 at the Université Virtuelle du Senegal (UVS) in hybrid mode. The technical program consisted of 11 full papers including an invited paper in oral presentation sessions at the main conference tracks. The conference tracks were as follows: Track 1 - Data Science and Artificial Intelligence; Track 2 - Telecom and Artificial Intelligence; and Track 3 - IoT and ICT Applications. In addition to the technical presentations, the technical program also featured two keynote speeches, one invited talk, three discussion panels, and one Young Researchers' session. The two keynote speakers were Guy Pujole (Université Sorbonne, France) who discussed the topic of "Network and Cloud Continuum" and Moulaye Akhloufi (University of Moncton, Canada) who gave a talk on "AI for computer vision and image analysis". The invited talk was about "Digitalization and Employability" and was delivered by Antoine Ngom, General Director of GSIE Technology and President of the Senegal Association of ICT Professionals. In the three discussion panels, attendees exchanged ideas on three important topics: cybersecurity, research funding in Africa, and ICT training curricula and jobs in the digital era. Finally, the technical program ended with a series of talks by Ph.D students during the Young Researchers' session. After two and half day of fruitful presentations, exchanges, and debates around technology and Africa, conference attendees were invited to relax with social events that included a gala dinner and a visit to "Monument de la Renaissance Africaine", one of the marvels of the splendid city of Dakar.

Coordination with ASCII's board members led by Bamba Gueye and its Research and Innovation Commission, led by Dame Diongue was essential for the success of the conference. We sincerely appreciate their constant support and guidance. It was also a great pleasure to work with such an excellent organizing committee team for their hard work in organizing and supporting the conference. In particular, we are grateful to the Technical Program Committee, who completed the peer-review process for the technical papers and contributed to a high-quality technical program. We are also grateful to Conference General Chairs El Hadji Mamadou Nguer and Mor Bakhoum for their support and to all the authors who submitted their papers to the CNRIA 2021 conference and workshops.

We strongly believe that the CNRIA conference provides a good forum for all researchers, developers, and practitioners to discuss all science and technology aspects that are relevant to ICT in Africa. We also expect that the future conferences will be as successful and stimulating as CNRIA 2021, as indicated by the contributions presented in this volume.

Youssou Faye
Assane Gueye
Bamba Gueye

Organization

Steering Committee

Youssou Faye	Université Assance Seck de Ziguinchor (UASZ), Sénégal
Assane Guèye	Carnegie Mellon University Africa (CMU-Africa), Rwanda
Bamba Gueye	Université Cheikh Anta Diop de Dakar, Sénégal
Dame Diongue	Université Gaston Berger, Sénégal

Organizing Committee

General Chair

Elhadji Mamadou Nguer	Université Virtuelle du Sénégal (UVS), Sénégal

General Co-chair

Mor Bakhoum	Université Virtuelle du Sénégal (UVS), Sénégal

Technical Program Committee Chair and Co-chair

Youssou Faye	Université Assane Seck de Ziguinchor (UASZ), Sénégal
Assane Gueye	Carnegie Mellon University Africa (CMU-Africa), Rwanda

Sponsorship and Exhibit Chair

Mayoro Cisse	Université Virtuelle du Sénégal (UVS), Sénégal

Local Chair

Elhadji Mamadou Nguer	Université Virtuelle du Sénégal (UVS), Sénégal

Publications Chair

Youssou Faye	Université Assane Seck de Ziguinchor (UASZ), Sénégal

Web Chairs

Cheikh Diawara	Université Virtuelle du Sénégal (UVS), Sénégal
Gaoussou Camar	Université Alioune Diop de Bambey, Sénégal

Demos Chairs

Khalifa Sylla	Université Virtuelle du Sénégal (UVS), Sénégal
Ndéye Massata Ndiaye	Université Virtuelle du Sénégal (UVS), Sénégal

Tutorials Chairs

Khalifa Sylla	Université Virtuelle du Sénégal (UVS), Sénégal
Ndéye Massata Ndiaye	Université Virtuelle du Sénégal (UVS), Sénégal

Technical Program Committee

Youssou Faye	Université Assane Seck de Ziguinchor, Sénégal
Assane Gueye	Carnegie Mellon University Africa (CMU-Africa), Rwanda
Bamba Gueye	Université Cheikh Anta Diop de Dakar, Sénégal
Cheikh Ba	Université Gaston Berger, Sénégal
Mahamadou Traore	Université Gaston Berger, Sénégal
Samba Sidibe	Ecole Polytechnique de Thies, Sénégal
Moussa Déthié Sarr	Université de Thies, Sénégal
Edouard Ngor Sarr	Université Assane Seck de Ziguinchor, Sénégal
Nathalie Pernelle	Université Sorbonne Paris Nord, France
Gorgoumack Sambe	Université Assane Seck de Ziguinchor, Sénégal
El Hadji Mamadou Nguer	Université Virtuelle du Sénégal, Sénégal
Ndeye Fatou Ngom	Ecole Polytechnique de Thies, Sénégal
Diery Ngom	Université Alioune Diop Gueye, Sénégal
El Hadji Malick Ndoye	Université Assane Seck de Ziguinchor, Sénégal
Djamal Benslimane	Université Claude Bernard Lyon 1, France
Ndeye Massata Ndiaye	Université Virtuelle du Sénégal, Sénégal
Marie Ndiaye	Université Assane Seck de Ziguinchor, Sénégal
Abdourahime Gaye	Université Alioune Diop de Bambey, Sénégal
Moindze Soidridine Moussa	Université Cheikh Anta Diop de Dakar, Sénégal
Ahmed Mohameden	Université de Nouakchott, Mauritania
Maissa Mbaye	Université Gaston Berger, Sénégal
Ladjel Bellatreche	ENSMA, France
Jean-Charles Boisson	Université de Reims Champagne-Ardenne, France
Gaoussou Camara	Université Alioune Diop de Bambey, Sénégal
Eddy Caron	ENS Lyon, France
Papa Alioune Cisse	Université Assane Seck de Ziguinchor, Sénégal

Serigne Diagne	Université Assane Seck de Ziguinchor, Sénégal
Cherif Diallo	Université Gaston Berger, Senegal
Mouhamadou Saliou Diallo	Université de Strasbourg, France
Moussa Diallo	Université Cheikh Anta Diop de Dakar, Sénégal
Ousmane Diallo	Université Assane Seck de Ziguinchor, Sénégal
Awa Diatara	Université Gaston Berger, Sénégal
Abel Diatta	Université Assane Seck de Ziguinchor, Sénégal
Samba Diaw	Université Cheikh Anta Diop de Dakar, Sénégal
Ousmane Dieng	Université Gaston Berger, Sénégal
Abdou Kâ Diongue	Université Gaston Berger, Sénégal
Dame Diongue	Université Gaston Berger, Sénégal
Abdoukhadre Diop	Université Alioune Diop de Bambey, Sénégal
Babacar Diop	Université Gaston Berger, Sénégal
Lamine Diop	Université Gaston Berger, Sénégal
Madiop Diouf	Université Assane Seck de Ziguinchor, Sénégal
Mamadou Diallo Diouf	Ecole Polytechnique de Thies, Sénégal
Ibra Dioum	Université Cheikh Anta Diop de Dakar, Sénégal
Khadim Drame	Université Assane Seck de Ziguinchor, Sénégal
Ibrahima Fall	Université Cheikh Anta Diop de Dakar, Sénégal
Lamine Faty	Université Assane Seck de Ziguinchor, Sénéga
André Faye	Université Gaston Berger, Sénégal
Issa Faye	Université Assane Seck de Ziguinchor, Sénégal
Olivier Flauzac	Université de Reims Champagne-Ardenne, France
Mouhamadou Gaye	Université Assane Seck de Ziguinchor, Sénégal
Arnaud Giacometti	Université de Tours, France
Ibrahima Gueye	Université Cheikh Anta Diop de Dakar, Sénégal
Mbaye Babacar Gueye	Université Cheikh Anta Diop de Dakar, Sénégal
Modou Gueye	Université Cheikh Anta Diop de Dakar, Sénégal
Youssou Kasse	Université Alioune Diop de Bambey, Sénégal
Dominique Laurent	CY Cergy Paris Université, France
Lemia Louail	University of Sétif 1, Algeria
Ahmath Bamba Mbacke	Université Cheikh Anta Diop de Dakar, Sénégal
Mouhamadou Lamine Ba	Université Alioune Diop de Bambey, Sénégal
Guy Mbatchou	Université Assane Seck de Ziguinchor, Sénégal

Contents

Data Science and Artificial Intelligence

A Distributed Memory-Based Minimization of Large-Scale Automata

Alpha Mouhamadou Diop[(⊠)] [iD] and Cheikh Ba[iD]

LANI - Université Gaston Berger, Saint-Louis, Senegal
{diop.alpha-mouhamadou,cheikh2.ba}@ugb.edu.sn

Abstract. Big data are data that are not able to fit in only one computer, or the calculations are not able to fit in a single computer memory, or may take impractical time. Big graphs or automata are not outdone. We describe a memory-based distributed solution of such large-scale automata minimisation. In opposition to previous solutions, founded on platforms such as MapReduce, there is a twofold contribution of our method: (1) a speedup of algorithms by the use of a memory-based system and (2) an intuitive and suitable data structure for graph representation that will greatly facilitate graph programming. A practical example is provided with details on execution. Finally, an analysis of complexity is provided. The present work is a first step of a long term objective that targets a advanced language for large graphs programming, in which the distributed aspect is hidden as well as possible.

Keywords: Big data · Distributed computing · Automata minimizing

1 Introduction

Automata are a mathematical model of computation, a simple and powerful abstract machine modeling many problems in computer science. They have been studied long before the 1970s. Automata are closed under several operations and this advantage makes them suited for a modular approach in many contexts. Applications include natural language processing, signal processing, web services choreography and orchestration, compilers, among so many others.

Determinization, intersection, and minimisation are some examples of problems linked to automata and that have been largely studied in the literature. Since one of our main objectives is to have artifacts for distributed programming, we choose to tackle the minimisation of large-scale or big automata. When it is about very large graphs (classical graphs or automata), distributed methods such as disk based parallel processing [22,23], MapReduce model [3,8,12,13,16,17] and distributed memory-based system [4,15,24] are considered.

Presently we study the innovative use of Pregel [19], a model for large-scale graph processing, to implement automata minimisation. Unlike the work we compare ourselves [12], we speed up the whole process by the use of a memory-based distributed system, and we don't need to use a counterintuitive data structure.

© ICST Institute for Computer Sciences, Social Informatics and Telecommunications Engineering 2021
Published by Springer Nature Switzerland AG 2021. All Rights Reserved
Y. Faye et al. (Eds.): CNRIA 2021, LNICST 400, pp. 3–14, 2021.
https://doi.org/10.1007/978-3-030-90556-9_1

Our long term goal is a high level language for distributed graph, a sort of language that hides, as well as possible, the distributed nature of the graph. In this way, the present work is one of the first steps, and it provides interesting new artefacts in addition the ones that appear in starting works [5,6].

The remaining of our paper is structured as follows: Sect. 2 presents some related works and Sect. 3 gives technical definitions and some backgrounds, that is automata and their minimisation in MapReduce, which is the work we compare ourselves. In Sect. 4 we describe our proposition, as well as some important discussions. Conclusions are shown in Sect. 5.

2   Related Works

Big data are collections of data that are too large − in relation to their processing − to be handled by classic tools. It concerns data collecting, storage, analysis, visualizing, querying and so on. Using a lot of computers for a shared storage and a parallel processing is a usual solution. Large scale graphs and automata are, of course, not outdone.

In this section we will point out two families of methods. The first one is about solutions that are built from-scratch, that is, not only the infrastructure (cluster of computers), but also its coordination. The second family concerns methods with a high level abstraction so that coordination complexity is hidden. In this way, users don't need low tasks programming skill and can only focus on functional aspects. We place ourselves in this second family of methods.

When it comes to an automaton, an important algorithm consist in obtaining its unique minimal and deterministic version. Some parallel algorithms consider shared RAM computers, using the EREW PRAM models [21] for instance. The aforementioned algorithms are applicable for a cluster of tightly coupled parallel computers with shared and heavy use of RAM. In addition, authors in [21] used a 512-processor CM-5 supermachine to minimise a *Deterministic Finite state Automaton* (DFA) with more than 500,000 states. In the case of a very large DFA, a disk storage will probably be needed. In this context, authors in [22,23] propose a parallel disk-based solution with a cluster of around thirty commodity machines to produce the unique and minimal DFA with a state reduction of more than 90%.

As presented above, the aforementioned solutions are part of the first family of methods.

Not long ago, Hadoop distributed platform [2] has been proposed by Google and has become rapidly the standard for big data processing. Its model for parallel programming is called MapReduce [9]. The objective of MapReduce is to facilitate parallel processing through only two routines: *map* and *reduce*. Data are randomly partitioned over computers and a parallel processing is done by executing the *map* and *reduce* routines on partitions. In Hadoop, different computers are connected in such a way that the complexity is hidden to end users, as if he is working with a single supercomputer. From that moment, several graph problems have been tackled by using MapReduce [3,8,16,17]: shortest path, graph twiddling, graph partitioning, minimum spanning trees, maximal matchings and so on. In the specific case of big automata, authors in [12,13] have proposed algorithms for NFAs intersection and DFA minimisation.

Even if MapReduce paradigm can be used for many graph concerns, it's well known that it is unsuitable for iterative graph algorithms. This a consequence of the immoderate input and output with HDFS, the Hadoop distributed file system, and data shuffling at all rounds. In this way, iterative algorithms can only be written by doing one job after another, because there is no natural support for iterative tasks in MapReduce. This often leads to considerable overhead. As a consequence, many graph processing frameworks using RAM are proposed: GraphLab [18], Spark/GraphX [11], PowerGraph [10], Google's Pregel [19] and Apache Giraph [1]. A vertex centric model is followed by the majority of these frameworks. For instance, in Pregel and Giraph frameworks, which are based on BSP [25], *Bulk Synchronous Parallel*, in each iteration, each graph node (vertex) may receive messages from some other ones, process a local task, and then may send messages to some other nodes.

Nonetheless, the particular case of automata is not very well exploited by the aforementioned frameworks. In this way, we propose the use of Pregel programming model to implement big DFA minimisation. Not only we enhance the process performance compared to the MapReduce solution in [12], but also we don't need to use and maintain their counterintuitive data structure.

3 Background and Terminology

3.1 Automata and Minimization

FSA, DFA and NFA: In this section we recall two kinds of automata or FSA (*Finite State Automata*), namely *Deterministic* and *Non-deterministic Finite state Automata*, respectively abbreviated by DFA and NFA.

A DFA consists in a finite set of states with labelled and directed edges between some pairs of states. Labels or letters come from a given alphabet. From each state, there is at most one edge labeled by a given letter. So, from a given state, a transition dictated by a given letter is *deterministic*. There is an *initial* or *start state* and also certain of the states are called *final* or *accepting*. A word w is accepted by the DFA if the letters of the word can be read through transitions from the start state to a accepting state. Formally, a DFA is a 5-tuple $A = (\Sigma, Q, q_s, \delta, F)$ such that the alphabet is Σ, the set of states is denoted by Q, the initial or start state is $q_s \in Q$, and the accepting or final states is the subset $F \subseteq Q$. The *transition function* is denoted by $\delta : Q \times \Sigma \rightarrow Q$, which decides, from a state and for a letter, in which state the system will move to.

An NFA is almost the same, with the difference that, from a state, we may have more than one edge with the same letter or label. The transition corresponding to the given letter is said to be *non-deterministic*. An NFA is formally defined by a 5-tuple $A = (\Sigma, Q, q_s, \delta, F)$. It differs from a DFA in the fact that $\delta : Q \times \Sigma \rightarrow 2^Q$, with 2^Q being the power set of Q. From a state and for a letter, an NFA can move to any one of the next states in a non-deterministic way.

The set of all words accepted by a FSA A defines the *language* accepted by A. It's denoted by $L(A)$. NFAs and DFAs are equivalent in terms of accepted languages. A language L^{reg} is *regular* if and only if there exists an FSA A such that $L^{reg} = L(A)$. Any NFA A can be transformed into a DFA A^D, such that $L(A^D) = L(A)$. This transformation is called *determinization*, and can be done by the *powerset construction* method. On the basis that a DFA transition function δ takes as arguments a state and a letter and returns a state, we denote by δ^* the extended transition function that takes a state p and a string $w = a_1 a_2 \cdots a_l$ and returns the unique state $q = \delta^*(p, w) = \delta(\delta(\cdots \delta(\delta(p, a_1), a_2) \cdots, a_{l-1}), a_l)$, which is the state that the automaton reaches when starting in state p and processing the sequence of symbols in w. Thus for a DFA A, the accepted language $L(A)$ can be defined as $\{w : \delta^*(q_s, w) \in F\}$. A DFA A is described as *minimal*, if all DFAs B that accept the same language $(L(A) = L(B))$ have at least as many states as A. There is a unique *minimal* and equivalent DFA for each regular language L^{reg}.

In Fig. 1 we depict two FSAs on $\Sigma = \{a, b\}$ and that accept words that start with "a" and end with "b": a NFA A (1-a) and its determinization A^D (1-b).

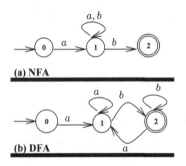

(a) NFA

(b) DFA

Fig. 1. An example of two FSAs.

Algorithm 1. Algorithm of Moore

Input: A DFA $A = (\{a_1, \cdots, a_k\}, Q, q_s, \delta, F)$
Output: $\pi = Q/_\equiv$
1: $i \leftarrow 0$
2: **for all** $p \in Q$ ▷ The initial partition
3:     **if** $p \in F$ **then**
4:         $\pi_p^i \leftarrow 1$
5:     **else**
6:         $\pi_p^i \leftarrow 0$
7:     **end if**
8: **end for**
9: **repeat**
10:     $i \leftarrow i + 1$
11:     **for all** $p \in Q$
12:         $\pi_p^i \leftarrow \pi_p^{i-1} \cdot \pi_{\delta(p,a_1)}^{i-1} \cdot \pi_{\delta(p,a_2)}^{i-1} \cdots \cdot \pi_{\delta(p,a_k)}^{i-1}$
13:     **end for**
14: **until** $|\pi^i| = |\pi^{i-1}|$

Minimization Algorithms: According to a taxonomy [26], the notion equivalent states is the one on which are based the majority of DFA minimisation algorithms. One exception is Brzozowski's algorithm [7] which is based on *determinization* and *reversal* of DFA. For a given DFA A, he showed that $(((A^R)^D)^R)^D$ is the minimal DFA for $L(A)$, knowing that the *reversal* of A is the NFA $A^R = (\Sigma, Q, F, \delta^R, \{q_s\})$, where $\delta^R = \{(p, a, q)$ such that $(q, a, p) \in \delta\}$. In other words, if $w = a_1 \cdots a_l \in L(A)$, then $w^R = a_l \cdots a_1 \in L(A^R)$. However it is quite clear that this method is costly.

The other algorithms, like Moore's [20] and Hopcroft's [14], are based on states equivalence. Two states p and q are said to be *equivalent*, wich is denoted by $p \equiv q$, if for all strings w, it holds that $\delta^*(p, w) \in F \Leftrightarrow \delta^*(q, w) \in F$. The relation $\equiv$ is an equivalence relation, and the induced partition of the state space Q is denoted by $Q/_\equiv$. The *quotient* DFA $A/_\equiv = (\Sigma, Q/_\equiv, q_s/_\equiv, \gamma, F/_\equiv)$, such that $\gamma(p/_\equiv, a) = \delta(p, a)/_\equiv$, is a *minimal* DFA, that is $L(A) = L(A/_\equiv)$. One important property is that this equivalence relation $\equiv$ can iteratively be computed as a sequence $\equiv_0, \equiv_1, \cdots, \equiv_m = \equiv$, such that $p \equiv_0 q$ if $p \in F \Leftrightarrow q \in F$, and $p \equiv_{i+1} q$ if $p \equiv_i q$ and $\forall a \in \Sigma, \delta(p, a) \equiv_i \delta(q, a)$. It is known that this sequence converges in at most l iterations, l being the length of the longest simple path from the initial state to any final state.

In the following, only one states equivalence-based solution will be considered. It is about Moore's algorithm [20]. Authors in [12] proposed Algorithm 1 as an implementation of Moore's algorithm. Algorithm 1 calculates the partition $Q/_\equiv$ by iteratively refining the initial partition $\pi = \{F, Q \setminus F\}$. At the end of the computation, $\pi = Q/_\equiv$. π is considered as a mapping that assigns a string of bits π_p — as a way to identify partition blocks — to each state $p \in Q$. The number of blocks in π is denoted by $|\pi|$, and the value of π at the ith iteration is denoted π^i.

3.2   DFA Minimization in MapReduce

Works in [12,13] are the unique contributions we know and that propose high level distributed platform to process automata related problems. In [12], authors describe MapReduce implementations of Moore's and Hopcroft's algorithms for DFA minimisation, as well as their analysis and experiments on several types of DFAs. As mentioned earlier, we will only focus on the algorithm of Moore, named *Moore-MR*.

We assume backgrounds on the MapReduce programming model [9], however we recall basic notions. This model is based on routines *map* and *reduce*, and the user has to implement them. The signature of `map` is $\langle K_1, V_1 \rangle \rightarrow \{\langle K_2, V_2 \rangle\}$ and the one of `reduce` is $\langle K_2, \{V_2\} \rangle \rightarrow \{\langle K_3, V_3 \rangle\}$. HDFS is used to store data, and each mapper task will handle a part of this input data. In each MapReduce round or iteration, all mappers emit a list of key-value couples $\langle K, V \rangle$. This list is then partitioned by the framework according to the values of K. All couples having the same value of K belong to the same partition $\langle K, [V_1, \cdots, V_l] \rangle$, which will be sent to the same reducer.

Moore-MR [12] consists in a preprocessing step, and several iterations of MapReduce.

Preprocessing: Since the algorithm is an repetitive refinement of the initial partition $\pi = \{F, Q \setminus F\}$, and due to the nature of MapReduce paradigm, authors have to build and maintain a set Δ from $A = (\{a_1, \cdots, a_k\}, Q, q_s, \delta, F)$. The set Δ consists in labeled transitions (p, a, q, π_p, D) such that: π_p is a string of bits representing the initial block of p, $D = +$ tells that the tuple represents a transition, whereas $D = -$ tells that the tuple is a "dummy" transition that hold in its fourth position information of the initial block of state q. In addition, tuples $(r_i, a_{i_j}, p, \pi_p, -)$ are also in Δ, because state p new block will be required in the following round when it's come to update states $r_1, \cdots, r_m$ block annotation, where $(r_1, a_{i_1}, p), \cdots, (r_m, a_{i_m}, p)$ are transitions that lead to p.

***map* Routine:** Authors in [12] define ν as the number of reducers, and $h :$ $Q \rightarrow \{0, \cdots, \nu - 1\}$ as a hash function. During iteration i, each mapper receives a part of Δ. For every tuple $(p, a, q, \pi_p^{i-1}, +)$, the mapper outputs key-value couple $\langle h(p), (p, a, q, \pi_p^{i-1}, +) \rangle$. And for every tuple $(p, a, q, \pi_p^{i-1}, -)$, the mapper outputs couples $\langle h(p), (p, a, q, \pi_p^{i-1}, -) \rangle$ and $\langle h(q), (p, a, q, \pi_p^{i-1}, -) \rangle$.

***reduce* Routine:** Every reducer $\rho \in \{0, \cdots, \nu - 1\}$ gets, for all $p \in Q$ such that $h(p) = \rho$, outgoing transitions $(p, a_1, q_1, \pi_p^{i-1}, +), \cdots, (p, a_k, q_k, \pi_p^{i-1}, +)$, along with "dummy" transitions $(p, a_1, q_1, \pi_{q_1}^{i-1}, -), \cdots, (p, a_k, q_k, \pi_{q_k}^{i-1}, -)$ and $(r_1, a_{i_1}, p, \pi_p^{i-1}, -), \cdots, (r_m, a_{i_m}, p, \pi_p^{i-1}, -)$.

The reducer is now able to compute $\pi_p^i \leftarrow \pi_p^{i-1} \cdot \pi_{q_1}^{i-1} \cdot \pi_{q_2}^{i-1} \cdot \ldots \cdot \pi_{q_k}^{i-1}$ and write the new value π_p^i in the tuples $(p, a, q_j, \pi_p^{i-1}, +)$, for $j \in \{1, \cdots, k\}$, and $(r_j, a, p, \pi_p^{i-1}, -)$, for $j \in \{1, \cdots, m\}$, which it then returns. The reducer may return a "change" tuple $(p, true)$ as well, if the new value of π_p^i signifies the increase of the number of blocks in π_i. In this case, the algorithm need another iteration of MapReduce.

It goes without saying that it would be better if one could not have to use a structure like Δ, which is not very intuitive. Fortunately, our proposition presented in Sect. 4 will do without any extra data structure, in addition to the speedup of the whole process.

4   Our Memory-Based Approach

4.1   Pregel System

Pregel [19] is one of the first BSP (Bulk Synchronous Parallel, [25]) implementations that provides an high level API for programming graph algorithms. BSP is a model for parallel programming, and that uses MPI (message passing interface). It has been developed for scalability by parallelizing tasks over multiple computers. Apache Giraph [1] is an open-source alternative. Pregel computing paradigm is said to be "think like a vertex" as long as graph processing is done in terms of what each graph node or vertex has to process. Edges are communication means between vertices. During a superstep, a vertex can run "`compute()`", a single routine coded by the user, exchange messages with any other vertex and may change its state (active/inactive). BSP is based on synchronization barrier (Fig. 2-**a**) to ensures that all messages sent in a previous superstep will be received in the following superstep. Function `voteToHalt()` can be called by a vertex in order to be inactive during the following superstep. But, this vertex will be active if it receives a message. The Pregel process will end if at the beginning of a superstep all vertices are inactive. Figure 2-**b** shows state transition of a vertex.

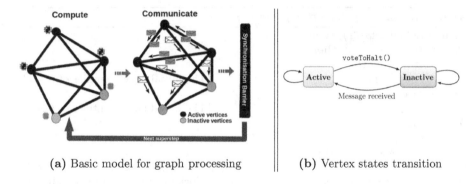

(a) Basic model for graph processing   **(b)** Vertex states transition

Fig. 2. About BSP model

4.2 Solution Inspired by the One of MapReduce

Given automaton $A = (\{a_1, \cdots, a_k\}, Q, q_s, \delta, F)$, we propose the `compute()` function (Algorithm 2) in order to minimise A. We remind that `compute()` will be executed by every active vertex during each superstep. And of course these vertices are distributed all over the memories of the distributed system (cluster of computers). Since the present work consider automata, Pregel's vertices represent automata states, and Pregel's edges represent automata transitions. Algorithm 2 consists in an alternation of two types of supersteps or rounds, thas is even $(0, 2, 4, \cdots)$ and odd $(1, 3, 5, \cdots)$ supersteps. New block identifiers are created in even supersteps. In order to create two rounds later the next block identifier of a state p, p has to ask for data to each of its target states q_i, to be sent in the next and odd round. Actually, odd rounds are dedicated to sending messages from target states q_i to soliciting state p.

At the start (line 2), that is the first superstep, all the automaton states are active and will divide up into the two initial blocks of the partition: the block identifier of a state p is 1 if p is final (line 4) and 0 otherwise (line 6). Apart from the first superstep, new block identifier π_p^{new} of a state p will be created depending on it previous block identifier π_p and the ones of its targets states π_{q_i} (line 13). For this purpose, state p asked for information, two steps earlier, from each target state q_i (line 16). Thus, this query is received and handled by each q_i in the previous (odd) round. In the end, p will get solicited data (through messages in M) in the present (even) round or superstep, and can extract block identifiers of each target state q_i (from line 9 to 12). The block identifier of p will be updated (line 14) in order to be sent, in the following (odd) superstep, to requesting states leading to p, that is, states for which p is a target state.

Finally, as said above, odd rounds are dedicated to providing data from target states q_i to each asking state p (line 20).

Concerning termination detection (line 24), we can be inspired by the MapReduce termination detection, or use the original Moore's algorithm termination detection. We'll consider the first one, and the second one will be discussed in Sect. 4.3.

Algorithm 2. *compute(vertex p, messages M)*

```
1:  if (EVEN_SUPERSTEP) then
2:      if (superstep = 0) then
3:          if (p ∈ F) then
4:              π_p ← 1                              {The initial bock identifier of p is 1}
5:          else
6:              π_p ← 0                              {The initial bock identifier of p is 0}
7:          end if
8:      else
9:          for each a_i ∈ Σ do
10:             q_i ← δ(p, a_i)                      {q_i is the target state of p related to a_i}
11:             π_{q_i} ← M.get_π(q_i)               {Get block identifiers of target states q_i}
12:         end for
13:         π_p^{new} ← π_p · π_{q_1} · π_{q_2} · ... · π_{q_k}    {The new bock identifier π_p^i of p}
14:         π_p ← π_p^{new}                          {Updating identifier for the next round}
15:     end if
16:     sendMessage(p.ID, q_i);        {Asks for data (π_{q_i}) from each state q_i = δ(p, a_i)}
17: else
18:     {//ODD_SUPERSTEP}
19:     for each p.ID in M do
20:         sendMessage(π_p, p.ID)                   {to send message (π_p) to soliciting p}
21:     end for
22: end if
23: if (NO_NEW_BLOCK) then
24:     p.voteToHalt();
25: end if
```

The block identification proposed in [12] is sophisticated enough to "locally" detect the appearance of a new block. In this way, from a block identifier π_p, we can know if a new block was created for p. The block π_p is a bit-string consisting of $k + 1$ of previous iteration. Having an increase of the different components signifies the creation of a new block. Algorithm 2 has to be completed by adding the following instruction: "if (NEW_BLOCK) then sendToAggregator ('change')" between lines 14 and 15. Pregel's *aggregators* enable "global" information exchange and will decide, in our case, to stop the whole process when no new block creation is detected, that is, the *aggregator* doesn't receive any "*change*" message from states. Example 1 gives running details of Algorithm 2 on a small automaton.

Example 1. Let us consider automaton $A = (\Sigma, Q, q_s, \delta, F)$, with $\Sigma = \{a, b\}$, $Q = \{0, 1, 2, 3\}$, $q_s = 0$, $F = \{3\}$ and δ presented in a state-transition table (Table 1-(a)) showing what states automaton A will move to, depending on present states and input symbols or letters.

In order to obtain A^{min}, the minimal version of automaton A, we give some details of the execution of Algorithm 2 in Table 1-(b). But only even supersteps are described. In superstep 0, states are divided up into the two initial blocks "0" and "1". State 0 belongs to block with identifier "0". Then, it will solicit data (block identifiers) from its target states 0 and 3. In superstep 1, these states will send their block identifiers to all asking states. For instance, state 3 will send identifier "1" to state 0. In next superstep (2), state 0 will receive requested data, create the new block identifier and issue a message to indicate that a new block is detected. This process will continue until no new block creation is detected by any of the vertices. This is what happens in superstep 6 where each block identifier π_i^6 has the same number of different components as its previous identifier π_i^4. Finally, we remark that $\pi_0 = \pi_2$, and the state-transition table of A^{min} is given in Table 1-(c). $\qquad\square$

Table 1. (a), (c) State-transition tables of A and A^{min}. (b) An example of a running.

(a)

δ	a	b
▶ 0	0	3
1	1	2
2	2	3
◀ 3	3	1

(c)

δ^{min}	a	b
▶ π_0	π_0	π_3
π_1	π_1	π_0
◀ π_3	π_3	π_1

(b)

Supersteps ⇓	π_0	π_1	π_2	π_3
0	$\pi_0^0 = 0$	$\pi_1^0 = 0$	$\pi_2^0 = 0$	$\pi_3^0 = 1$
2	$\pi_0^2 = 0.0.1$ Issues 'change'	$\pi_1^2 = 0.0.0$	$\pi_2^2 = 0.0.1$ Issues 'change'	$\pi_3^2 = 1.1.0$ Issues 'change'
4	$\pi_0^4 = 001.001.110$	$\pi_1^4 = 000.000.001$ Issues 'change'	$\pi_2^4 = 001.001.110$	$\pi_3^4 = 110.110.000$
6	$\pi_0^6 = 001001110.$ 001001110. 110110000	$\pi_1^6 = 000000001.$ 000000001. 001001110	$\pi_2^6 = 001001110.$ 001001110. 110110000	$\pi_3^6 = 110110000.$ 110110000. 000000001

4.3 Discussions

Termination Detection: As said in the previous section, we can be take our inspiration from the MapReduce termination detection, or use the original Moore's algorithm termination detection. The first kind of detection is used in Example 1. In fact, block identification proposed in [12] is sophisticated enough to "locally" detect the appearance of a new block, that is, from a local perspective or view. However it is not the case for the original termination detection for which we have to check if the number of blocks has changed from one round to another. If authors in [12] didn't find this kind of suitable block identification, they would be obliged to "globally" count, at each MapReduce round, the number of blocks. And for this purpose, and due to the nature of MapReduce paradigm, an extra round would be necessary after each functional or classic round.

Our solution (Algorithm 2) would not suffer from this concern since each vertex p just has to send its block identifier π_p to the *aggregator*; the latter will "globally" count the numbers of different block identifiers and decide whether the whole process will continue or not.

Comparative Complexities: In this section, an overview is given on complexities of the two different solutions.

Given a DFA A with n states, it is accepted that $\equiv = \equiv_{n-2}$ in the worst case [20], and thus, *Moore-MR* algorithm needs $(n-1)$ iterations in the worst case. Our Pregel solution needs $2 \times (n-1)$ supersteps. This is due to our odd supersteps devoted to sending data. Despite that, our solution is far and away faster then *Moore-MR* considering speed ratio between RAM and disk accesses. As a reminder, MapReduce model suffers from excessive input/output with HDFS (disk) and *"shuffle & sort"* at every iteration, whereas Pregel is memory-based.

Concerning communication cost, the one for *Moore-MR* is $O(k^2 n^2 \log n)$ [12], specially due to their set Δ they have to maintain. Contrary to them, in each superstep, our solution causes $k \times n$ messages sending - when vertices ask for information or when data are delivered to requesting states - and the n messages sent by vertices to *aggregator*. We therefore have a cost of $O(kn^2)$ in all.

Generally, Pregel solution costs the product of the number of rounds or supersteps and the cost of a superstep. The latter is the sum of three costs: the longest execution of `compute()` function, the maximum exchange of messages, and the barrier of synchronization.

Implementation: We use Apache Giraph [1] to implement our solution. It runs on Hadoop platform and thus uses HDFS to read input data and write transformed graph (or output). Input is composed of two parts: a directed graph and a the "compute()" function.

We use a custom class to handle a custom read of the input graph (files) from the file system (HDFS). In these files, each vertex is represented in one line with the following structure:

```
[vrtxID, vrtxData, [[destID_1, edgeVal_1], ..., [destID_k, edgeVal_k]]]
```

vrtxID is the vertex ID representing a state. vrtxData is a data structure that will contain the initial and next block identifiers of the considered state. destID_1 is a destination or target state ID, and edgeVal_1 is the corresponding edge value, that is, the symbol of the corresponding transition.

At the end of the computation, the resulting automaton can be obtained by changing Q by π (the set of all blocks), q_s by π_{q_s}, F by $\{\pi_f \mid f \in F\}$ and each transition $(p, a, q) \in \delta$ by transition (π_p, a, π_q).

Perspectives: As long as our long term objectives is to find relevant programming artifacts, for a high level language for distributed graph, and with which and the distributed aspects are hidden at most, we have to find a distributed graph object, general enough to allow us to faithfully and automatically transcribe non-distributed graph algorithms. Clearly, when we consider the original Moore's algorithm, it starts with two initial blocks, and these blocks are refined (broken into sub-blocks) at every round. Unfortunately, this block object cannot be represented as a single vertex since it may not fit in a single computer RAM (one initial block can contains all the automaton vertices, for instance if all of them are final). A future work will consist of defining a logical structure over the vertices, a kind of a graph distributed object, composed of several vertices and that may have its "`compute()`"-like function.

5   Conclusion

In this proposition, we have described and implemented a solution based on a cluster distributed memory to process big DFA minimisation. In fact, our proposition is based a model for big graph processing, namely Pregel/Giraph. Unlike the work we compare ourselves, we speed up the whole process by the use of a memory-based distributed system, and we don't need to use and maintain a counterintuitive data structure. In fact, Pregel offers an intuitive and suitable data structure for graph representation that greatly facilitates graph programming. A running example is given, as well as details on execution and complexity analysis. Finally, some important future works are described.

References

1. The Apache Software Foundation: Apache giraph. https://giraph.apache.org/
2. The Apache Software Foundation: Apache hadoop. https://hadoop.apache.org/
3. Aridhi, S., Lacomme, P., Ren, L., Vincent, B.: A mapreduce-based approach for shortest path problem in large-scale networks. Eng. Appl. Artif. Intell. **41**, 151–165 (2015)
4. Aridhi, S., Montresor, A., Velegrakis, Y.: BLADYG: a graph processing framework for large dynamic graphs. Big Data Res. **9**, 9–17 (2017)
5. Ba, C., Gueye, A.: On the distributed determinization of large NFAs. In: 2020 IEEE 14th International Conference on Application of Information and Communication Technologies (AICT), pp. 1–6, October 2020
6. Ba, C., Gueye, A.: A BSP based approach for NFAs intersection. In: Qiu, M. (ed.) ICA3PP 2020. LNCS, vol. 12452, pp. 344–354. Springer, Cham (2020). https://doi.org/10.1007/978-3-030-60245-1_24
7. Brzozowski, J.: Canonical regular expressions and minimal state graphs for definite events (1962)
8. Cohen, J.: Graph twiddling in a mapreduce world. Comput. Sci. Eng. **11**(4), 29–41 (2009)
9. Dean, J., Ghemawat, S.: MapReduce: simplified data processing on large clusters. Commun. ACM **51**(1), 107–113 (2008)
10. Gonzalez, J.E., Low, Y., Gu, H., Bickson, D., Guestrin, C.: Powergraph: distributed graph-parallel computation on natural graphs. In: 10th USENIX OSDI 2012, Hollywood, CA, USA, 8–10 October 2012, pp. 17–30 (2012)
11. Gonzalez, J.E., Xin, R.S., Dave, A., Crankshaw, D., Franklin, M.J., Stoica, I.: Graphx: Graph processing in a distributed dataflow framework. In: 11th USENIX OSDI 2014, Broomfield, CO, USA, 6–8 October 2014, pp. 599–613 (2014)
12. Grahne, G., Harrafi, S., Hedayati, I., Moallemi, A.: DFA minimization in mapreduce. In: Afrati, F.N., Sroka, J., Hidders, J. (eds.) Proceedings of the 3rd ACM SIGMOD Workshop on Algorithms and Systems for MapReduce and Beyond, BeyondMR@SIGMOD 2016, San Francisco, CA, USA, 1 July 2016, p. 4. ACM (2016)

13. Grahne, G., Harrafi, S., Moallemi, A., Onet, A.: Computing NFA intersections in map-reduce. In: Proceedings of the Workshops of the EDBT/ICDT 2015 Joint Conference, Brussels, Belgium, 27 March 2015. CEUR Workshop Proceedings, vol. 1330, pp. 42–45 (2015)

14. Hopcroft, J.E.: An n log n algorithm for minimizing states in a finite automaton. Technical Report, Stanford, CA, USA (1971)

15. Ko, S., Han, W.: Turbograph++: a scalable and fast graph analytics system. In: Das, G., Jermaine, C.M., Bernstein, P.A. (eds.) Proceedings of the 2018 International Conference on Management of Data, SIGMOD Conference 2018, Houston, TX, USA, 10–15 June 2018, pp. 395–410. ACM (2018)

16. Lattanzi, S., Mirrokni, V.S.: Distributed graph algorithmics: theory and practice. In: WSDM, pp. 419–420 (2015). http://dl.acm.org/citation.cfm?id=2697043

17. Lattanzi, S., Moseley, B., Suri, S., Vassilvitskii, S.: Filtering: a method for solving graph problems in mapreduce. In: Rajaraman, R., auf der Heide, F.M. (eds.) SPAA 2011: Proceedings of the 23rd Annual ACM Symposium on Parallelism in Algorithms and Architectures, San Jose, CA, USA, 4–6 June 2011 (Co-located with FCRC 2011), pp. 85–94. ACM (2011)

18. Low, Y., Gonzalez, J., Kyrola, A., Bickson, D., Guestrin, C., Hellerstein, J.M.: Distributed graphlab: a framework for machine learning in the cloud. PVLDB 5(8), 716–727 (2012)

19. Malewicz, G., et al.: Pregel: a system for large-scale graph processing. In: Proceedings of the ACM SIGMOD International Conference on Management of Data, SIGMOD 2010, Indianapolis, Indiana, USA, 6–10 June 2010, pp. 135–146. ACM (2010)

20. Moore, E.F.: Gedanken-experiments on sequential machines. In: Shannon, C., McCarthy, J. (eds.) Automata Studies, pp. 129–153. Princeton University Press, Princeton, NJ (1956)

21. Ravikumar, B., Xiong, X.: A parallel algorithm for minimization of finite automata. In: Proceedings of IPPS 1996, The 10th International Parallel Processing Symposium, 15–19 April 1996, Honolulu, USA, pp. 187–191. IEEE Computer Society (1996)

22. Slavici, V., Kunkle, D., Cooperman, G., Linton, S.: Finding the minimal DFA of very large finite state automata with an application to token passing networks. CoRR abs/1103.5736 (2011)

23. Slavici, V., Kunkle, D., Cooperman, G., Linton, S.: An efficient programming model for memory-intensive recursive algorithms using parallel disks. In: International Symposium on Symbolic and Algebraic Computation, ISSAC 2012, Grenoble, France - 22–25 July 2012, pp. 327–334. ACM (2012)

24. Su, J., Chen, Q., Wang, Z., Ahmed, M.H.M., Li, Z.: Graphu: a unified vertex-centric parallel graph processing platform. In: 38th IEEE International Conference on Distributed Computing Systems, ICDCS 2018, Vienna, Austria, 2–6 July 2018, pp. 1533–1536. IEEE Computer Society (2018)

25. Valiant, L.G.: A bridging model for parallel computation. Commun. ACM 33(8), 103–111 (1990)

26. Watson, B.: A taxonomy of finite automata minimization algorithms (1993)

Short-Term Holt Smoothing Prediction Model of Daily COVID-19 Reported Cumulative Cases

Ousseynou Mbaye[1(✉)], Siriman Konare[2], Mouhamadou Lamine Ba[1],
and Aba Diop[1]

[1] Université Alioune Diop de Bambey, Bambey, Senegal
{ousseynou.mbaye,mouhamadoulamine.ba,aba.diop}@uadb.edu.sn
[2] Université Gaston Berger de Saint Louis, Saint-Louis, Senegal
siriman.konare@ugb.edu.sn

Abstract. COVID-19 is the most deadly respiratory diseases worldwide known so far. It is a real public health problem against which contingency measures such as social distancing and lock down are often used to decrease the number of cases when it increases exponentially. These measures along with their impacts are set based on knowledge about the propagation of the disease, in particular the daily reported new and total cases within a given country. To plan in advance efficient contingency measures in order to stop its rapid propagation and to mitigate a possible explosion of the active cases leading to an uncontrolled situation and a saturation of health structures, governments need to have an indication about the potential number of total cases during incoming days; prediction models such as SIR algorithm try to provide such a kind of prediction. However, 'existing models like SIR are complex and consider many unrealistic parameters. This paper proposes, based on Holt's smoothing method combined with a logarithmic function for cold start, a very simple short-term prediction of the daily total number of COVID-19 cases. Our experimental evaluation over various COVID-19 real-world datasets from different countries show that our model, particularly using a linear trend function, gives results with low error rates. We also show that our approach can be generalized to all countries around the world.

Keywords: COVID-19 · Total cases · Prediction model holt smoothing · Trend · Evaluation performance

1 Introduction

In December 2019 [7], a cluster of pneumonia cases due to a newly identified coronavirus, referred to as COVID-19, has been discovered in Wuhan, a city in China. At that time, such an unknown disease showed a high human transmission rate leading to a rapid global propagation so that most of the countries worldwide have reported local or imported new cases within a very short time period.

© ICST Institute for Computer Sciences, Social Informatics and Telecommunications Engineering 2021
Published by Springer Nature Switzerland AG 2021. All Rights Reserved
Y. Faye et al. (Eds.): CNRIA 2021, LNICST 400, pp. 15–27, 2021.
https://doi.org/10.1007/978-3-030-90556-9_2

In few months, various hot spots have quickly appeared in Asia, Europe, America, and Africa pushing the experts of the World Health Organization (WHO) to declare COVID-19 as pandemic. Meanwhile they recommended governments to set contingency measures such as social distancing, systematic wearing of medical masks by citizens, mobility reduction or lock down in order to stop the increasing expansion of the virus. This rapid spread has been accompanied with thousands of deaths worldwide caused by the severe form of the disease with a high saturation of health structures in European countries in particular.

WHO has officially adopted Coronavirus 2019 (COVID-19 in short) as the official naming of this new respiratory disease in February 11, 2020 while at the same time the Coronavirus Study Group, an international committee that gathers several renowned health experts, has proposed a more scientific name which is SARS-CoV-2. This study aims at providing a prediction model of the daily total number of reported cases in a given country which information is critical in order to fight against the spread of the disease with suitable actions or to evaluate the impacts of already taken contingency measures. The number of new or cumulative total cases per day is one of the most important indicator that is monitored by governments and health organizations when dealing with pandemic disease. Till now, COVID-19 crisis is still ongoing despite the development of promising vaccines and still impacts highly the citizens of each country in all aspects of their daily life. Adaptability, psychological risk, social interactions, daily usual activities or projection in the future, economic activities are challenging for everyone. Various contingency measures such as social distancing, confinement and curfew are often used by governments to decrease the number of cases when it increases exponentially. These measures along with their impacts are set based on knowledge about the propagation of the disease, in particular the daily reported new and total cases within a given country. To plan in advance efficient contingency measures in order to stop its rapid propagation and to mitigate a possible explosion of the new cases leading to an uncontrolled situation and a saturation of health structures, governments need to have an indication about the potential number of new cases during incoming days. Several models have been proposed by researchers and scientists to describe the spread, the mode of contamination, the speed of contamination but also to trace back the movements of the population in order to easily find the persons in contact with the confirmed patients. However, 'existing prediction models, e.g. SIR model [6], which predict the number of new or total cases are complex and integrates sometimes unrealistic parameters.

The data about the total number of confirmed cases in a given country form a time series data and a prediction should be suitable for this kind of data. As a result, we introduce in this paper, based on Holt's smoothing method combined with a logarithm function for cold start, a very simple short-term prediction of the cumulative total number of confirmed COVID-19 cases every day. This also enables to deduce the number of daily new cases. The Holt algorithm is an extension of the exponential smoothing model for time series data with trends. In the case of COVID-19 the number of cases can increase or decrease depending on the country, the time period, the respect of social distance by citizens, etc.

To capture such an unknown evolution trend for a given country we evaluate with three different functions (linear, exponential and dampened) in our Holt's smoothing prediction approach. For the experimental evaluation we consider various real-world datasets from Senegal, Mali, France, USA, Australian, and Brazil. The choice of countries is far from being random. Indeed, the study is part of Senegal before being generalized to other countries such as Mali chosen on the basis of its proximity to Senegal. It was also necessary to be interested in what is happening in the European continent strongly impacted by COVID19, hence the choice of France. In addition the study was applied to the United States and Brazil to see its behavior in the American continent but also in Australia. As performance metrics, we measure the mean square error and mean absolute error of our model. Such an experimental evaluation over various COVID-19 real-world datasets from different countries show that our model, particularly using linear trend, gives results with low error rates. We also show that our approach can be generalized to all countries around the world.

The remaining of this study is organized as follows. First we review the related work in Sect. 2. We then introduce mathematical tools and methods used throughout this paper to build our proposed model in Sect. 3. Section 4 details the formalism of our prediction model. We present the results of the performance evaluation of our approach conducted on real-world datasets in Sect. 5 before concluding in Sect. 6 with some further research perspectives.

2   Related Work

Since the confirmation of the first case of COVID-19 in Senegal on March 02, 2020 the number of infected people continues to increase day by day. The spread of the virus is very dangerous and requires a number of measures to be taken. It is therefore very important to anticipate confirmed cases during incoming days to implement suitable protection plans. To gain better visibility into the spread of the virus, many studies have been done to predict the number of cases or deaths of COVID-19. It is in this sense that Zhao et al. [9] have proposed a mathematical model to estimate the actual number of officially unreported COVID-19 cases in China in the first half of January 2020. They concluded that there were 469 unreported cases between January 1 and January 15, 2020. Karako K et al. [5] have developed a stochastic transmission model by extending the SIR (Susceptible-Infected-Removed) epidemiological model with additional modeling of the individual action on the probability of staying away from crowded areas. In Iran, Zareie B et al. [8] have also used the SIR epidemiological model to estimate the number of COVID-19 cases. The analysis has been done on data between January 22 and March 24, 2020 and the prediction was made until April 15, 2020. The authors have come to the conclusion that approximately 29,000 people will be infected between March 25 and April 15, 2020. In Senegal, the authors of [6] have proposed a SIR epidemiological model combined with machine learning models to predict the evolution of the disease. Their results predicted the end of the pandemic in many countries by April 2020 at the latest.

Time series are also often used in disease prediction tasks. Indeed, the authors of [4] used prophet to predict the number of COVID-19 cases in India. They have observed that their fitted model is accurate within a certain range, and extreme prevention and control measures have been suggested in an effort to avoid such a situation. Alexandre Medeiros et al. [4] proposed a time series modified to study the incidence of the disease on mortality. At last, the authors of [2], in their COVID prediction, have studied the combination of time series with neural networks through the LSTM algorithm.

3   Background

This sections introduces the definition of the concepts underlying our proposed prediction approach. We start by the exponential smoothing function.

3.1   Simple Exponential Smoothing

Exponential smoothing [3] is a rather very simple prediction technique that tries to infer the value at $t+1$ from historical data. It applies to time series without trend[1]. The intuition is to give more importance to the last observations. In other words, the more recent the observation is, the greater is the weight that is associated to it. For example, it makes sense to assign higher weights to observations made yesterday than to observations made seven days ago. We do not extend a series as we would like, for instance as we can do with a simple regression. However, we try to obtain a smoothed value at t to simply transfer it to $t + 1$.

Let us define the notion of time series as follows.

Definition 1. *A time series is a series of data points indexed (or listed or graphed) in time order, i.e., a sequence taken at successive equally spaced points in time.*

Then, the exponential smoothing can be defined informally as follows.

Definition 2. *Exponential smoothing is a rule of thumb technique for smoothing time series data using the exponential window function.*

The aim of smoothing is to give a general idea of relatively slow changes of value with little attention paid to the close matching of data values, while curve fitting concentrates on achieving as close a match as possible. The exponential smoothing has only one component called *level* with a smoothing parameter denoted by α. It is formally defined as a weighted average of the previous level and the current observation.

$$y_{t+1} = \alpha \times y_t + \alpha(1 - \alpha) \times y_{t-1} + \alpha(1 - \alpha)^2 \times y_{t-2} + \ldots + \alpha(1 - \alpha)^n \times y_1 \quad (1)$$

where $0 \leq \alpha \leq 1$ is the smoothing parameter. The rate of weight reduction is controlled thanks to the smoothing parameter α. If α is large, more weight is given to more recent observations. We have the two following extreme cases.

---

[1] https://zhenye-na.github.io/2019/06/19/time-series-forecasting-explained.html.

- If $\alpha = 0$ we obtain the *average method* which corresponds to the case where the prediction of all future values is equal to the average of the historical data.
- If $\alpha = 1$ we obtain the *naive method* that just set all predictions to the value of the last observation.

3.2   Smoothing Holt's Method

Holt [1] is an extension of the simple exponential smoothing to allow the predictions of time series data with trends. The time series data with trend is informally defined as follows.

Definition 3. *A time series with trend is a set of ordered data points in which there is a long-term increase or decrease in the data. Such a trend does not have to be necessarily linear.*

Holt's method is based on a prediction function and two smoothing functions whose formalism are given below.

Prediction Function.

$$y_{t+h} = l_t + h \times b_t \tag{2}$$

Smoothing Level Function.

$$l_t = \alpha \times y_t + (1 - \alpha) \times (l_{t-1} + b_{t-1}) \tag{3}$$

Smoothing Trend Function.

$$b_t = \beta \times (l_t - l_{t-1}) + (1 - \beta) \times b_{t-1} \tag{4}$$

In the functions above, α ad β have the following semantics.

- $0 \le \alpha \le 1$ is the exponential smoothing parameter.
- $0 \le \beta \le 1$ is the smoothing parameter of the trend.

4   Short-Term Holt Smoothing Based Prediction Function

This sections presents the formalism of the model we propose for the prediction of the total number of reported cases in a given day. We consider the set of total number of reported COVID-19 in a country in a period of time as time series data and we propose a Holt's smoothing based prediction model as we will detail it next. The proposed model based on the Holt smoothing is a chronological approach whose different steps are sketched in Fig. 1 for the training, validation and prediction tasks.

Let us assume the time series data $X = (t_i, y_i)_{1 \le i \le n}$ for n natural numbers and a function $f : N^* \mapsto R^+$ defined by

$$f(t) = b \times t \times ln(1 + at) \tag{5}$$

where t denotes the rank of the date considered to the index i, ln corresponds to the logarithm function and y_i denote the number of COVID19 cases at date t_i. The coefficients a and b of the function f satisfy the following conditions:

- $0 \leq a \leq 1$; and
- $0 \leq b \leq 1$.

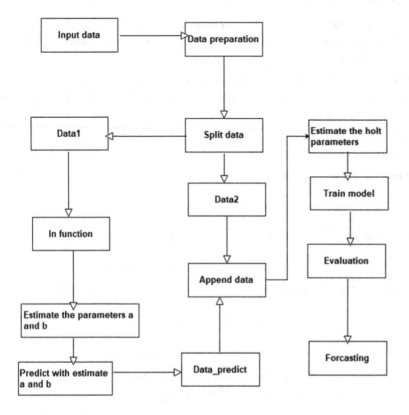

Fig. 1. Training, evaluation and prediction steps of our model

These coefficients have to be estimated from the real data. Then, for any natural number i, $0 \leq i \leq k$, (with k a natural number) we define a new series X_i^k as follows:

$$X_i^k = (t_i, f(t_i)) \tag{6}$$

In order to solve our prediction problem using Holt's method, we introduce a new series, denoted by X_2, obtained from the series $X = (t_i, y_i)_{1 \leq i \leq n}$ and $X_i^k = (t_i, f(t_i))$ in the following manner:

$$X_2 = \{(t_1, f(t_1)), (t_2, f(t_2)), (t_3, f(t_3)), \ldots, (t_k, f(t_k)), (t_1, y_1), \ldots, (t_n, y_n)\} \tag{7}$$

Let us set $Z_i = f(t_i)$ for $1 \leq i \leq k$ and $Z_i = X_i$ for $k+1 \leq i \leq n$ $(k \leq n)$, we simplify Eq. 7 in order to obtain the corresponding time series

$$Z = (t_i, Z_i)_{1 \leq i \leq n} \qquad (8)$$

on which we will apply the different trend functions (linear, exponential and dampened) based on the proposed Holt method with trend. We evaluate different trend functions because of the fact that we have no prior knowledge about the actual trend of the evolution of the total number of daily reported cases in a given country. Testing with several functions will help us to choose the optimal one according to the data.

5   Experimentation Evaluation

This section presents the results of our experimental evaluation of the our proposed Holt smoothing based prediction model of daily cumulative cases of COVID-19. Recall that we combine Holt with a logarithm function for solving the cold start problem. We measure the prediction errors of our model with the linear, dampened and exponential trends.

We start by presenting the setting up of our tests with the description of the various datasets used for the experiments, considered performance metrics in order to measure the efficiency of our model and the turning of the optimal values of the parameters of the model. We have implemented our proposed Holt prediction model with the different trend functions using Python programming language. All the implemented source codes as well used datasets are available at: https://github.com/siriman/covid-19-senegal.

5.1   Setting Up Our Experimentation

We first describe used real-world datasets and then we present the performance metrics that have been considered to measure the performance of our model. We end this section by presenting the choice of the optimal values of the input parameters.

Description of the Real-World Datasets. Numerous datasets have been used to evaluate the performance of our model. They come from authoritative sources, mainly from daily reports of each country. We have considered datasets about the cumulative reported cases at each date in Senegal, Mali, France, Australia, USA, and Brazil. Table 1 summarizes the statistics about those datasets by specifying the total cases, the time period of data collection, and the average number of cases per day for each country. For the specific case of Senegal, Fig. 2 shows collected time series data that have been reported from 02/03/2020 to 26/02/2021; the total reported cases for the last ten days and the shape of the curve of evolution of the reported total number of cases given the entire time period. One has to note that, however, these data may be biased due to the

Table 1. Statistics of collected datasets about five countries

Country	Time_Period	Total_Cases	Avg_cases
Senegal	2020-03-02 to 2021-02-26	34031	94
Mali	2020-01-24 to 2021-02-26	8358	24
France	2020-01-24 to 2021-02-26	3746707	9366
Australia	2020-01-24 to 2021-02-26	28965	72
USA	2020-01-24 to 2021-02-26	28486394	70861
Brazil	2020-01-24 to 2021-02-26	10455630	28489

fact that in countries like Senegal there is no massive screening of the population, i.e. only persons with symptoms have been tested to verify whether or not they suffer from the disease. For turning the parameters of the model and performing the validation of its performance, we have first focused on data from Senegal. Then, we have demonstrated the robustness and the generality of our approach to other countries such as Mali, France, USA, Australia, and Brazil. For the tests, we have considered 90% of each dataset for the training and the 10% remaining for the testing.

	Date	Number of Cases
352	2021-02-17	31771
353	2021-02-18	32099
354	2021-02-19	32378
355	2021-02-20	32630
356	2021-02-21	32927
357	2021-02-22	33099
358	2021-02-23	33242
359	2021-02-24	33453
360	2021-02-25	33741
361	2021-02-26	34031

(a) Last 10 days

(b) All days

Fig. 2. Cumulative number of reported cases in Senegal

Performance Metrics. After having trained our chronological model, we have then proceeded to its evaluation. In the literature, there are several existing performance measures intended to evaluate the accuracy of a given prediction model[2]. In this study we rely on the *Mean Square Error* and the *Mean Absolute Error* that are two popular metrics to measure the error rate of a chronological

---

[2] https://www.analyticsvidhya.com/blog/2019/08/11-important-model-evaluation-error-metrics/.

model. We provide their definitions below. Let Z_i be the real number of total cases at date i and P_i the predicted number of total cases.

- **Mean Square Error (MSE)** measures the average of the squares of the errors, that is, the average squared difference between the estimated values and the real values.

$$MSE = \frac{1}{n} \sum_{i=1}^{n} (Z_i - P_i)^2 \tag{9}$$

- **Mean Absolute Error (MAE)** corresponds to the arithmetic average of the absolute errors between real values and predicted values.

$$MAE = \frac{1}{n} \sum_{i=1}^{n} |Z_i - P_i| \tag{10}$$

We compare the MSE and MAE values of the three different trend functions (linear, dampened and exponential) when used with the Holt Smoothing model. The best prediction model corresponds to the one which presents the lowest values of MSE and MAE. We will show that our model present very low MSE and MAE rate errors in predicting the number of cumulative cases every day, in particular when used with the linear trend function.

5.2   Parameters Turning

The choice of the optimal values of the parameters α and β are critical for the performance of the Holt algorithm. Such a choice of these parameters are generally very subjective and might vary depending on the context of the study or the type of desired prediction. In this study, the parameters a and b for the logarithmic part of the model as well as the α and β coefficients of the **Holt** method are estimated for values between 0 and 1 with a step of 0.1 based on given training data of the Senegal's dataset.

Figures 3a, 3b, 3c and 3d give for each of these couples the variations of the MSE. Thus, we observe that the best pair of values of (a, b) is $(0.2, 0.7)$ with an MSE equals to 0.835. Meanwhile, the values of α and β are $(0.9, 0.3)$ with an MSE of 1634 for the linear trend, 1661 for the damped trend and of 1694 for the exponential trend. We choose these values of a and b as well as those of α and β that minimize the MSE for the rest of this evaluation.

5.3   Description and Analysis of the Obtained Results

In this section, we present the results of the prediction power of our model with each of the different trend functions. We calculate for each variant its **MSE** and **MAE**.

Table 2 presents the values of the MSE and the MAE of all the variants of the modified Holt's method. The curves in Fig. 4 compares the actual reported values and the predicted values about the cumulative number of COVID-19 cases

Table 2. Measures of the performance of the Holt model with various trend functions

Model	MSE	MAE
Holt with linear trend	1634.66	25.93
Holt with dumped trend	1661.52	26.29
Holt with exponential trend	1694.13	26.50

in Senegal for the linear, dampened and exponential trend functions respectively. The evolution of cumulative reported cases is given by the blue line while the cumulative number of cases predicted by our model is given by the red line. An analysis of the obtained results from the experiments of our Holt approach with the three trend functions show that all of them are quite accurate for the task of predicting the daily cumulative number of cases of COVID-19 in Senegal. However, in term of MSE and MAE we observe that combining Holt with the linear trend outperforms the others significantly. Indeed, such a combination provides the smallest MSE and MAE values which are respectively 1634.66 and 25.93. This indicates that our prediction differs, on average, by approximately

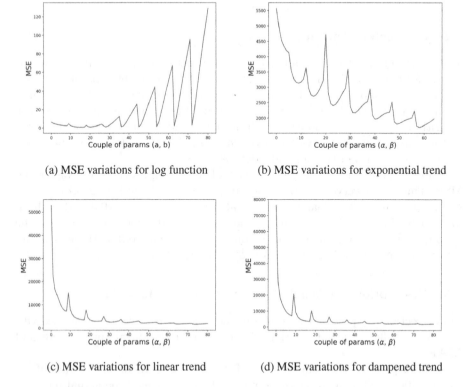

(a) MSE variations for log function

(b) MSE variations for exponential trend

(c) MSE variations for linear trend

(d) MSE variations for dampened trend

Fig. 3. Estimation of the optimal values of the parameters of our prediction model

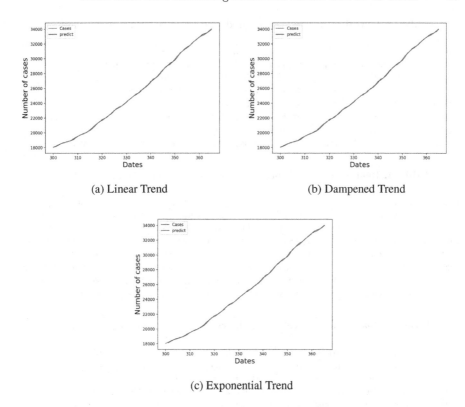

(a) Linear Trend   (b) Dampened Trend

(c) Exponential Trend

Fig. 4. Number of real cases vs. number of predicted cases

1634.66 from the actual data. In other words, this represents a difference of approximately 25.93 in absolute value from the actual data.

Unlike SIR (Susceptible, Infectious, or Recovered) models which are very complex and which involve a number of parameters that are very difficult to control and whose goal is to predict the basic reproduction number, R_0, our model which makes it possible to predict the number of cases of contamination in the near future can be of a capital contribution in the definition of policies for the fight against the pandemic.

We also evaluated and showed the robustness of our prediction model by testing it on datasets from Mali, France, Australia, Brazil, and USA. To this end, for each of these countries we have compared the real cumulative cases during the last five days with the values predicted by our model with the different trend functions in the same period. Tables 3, 4 and 5 contain the results of the comparison and show that Holt with the different trend functions capture well the evolution of the cumulative number of reported cases in these countries regarding the prediction error when comparing real and predicted values, e.g. the maximum deviation is equal to 8 for Mali.

Table 3. Real values vs. Values predicted by Holt with Linear trend

Date	Mali		France		Australia		USA		Brazil	
	Real values	Predicted values	Real values	Predicted values	Real values	Predicted values	Real values	Predicted values	Real values	Predicted
2021-02-22	8306	8311	3669354	3685625	28937	28934	28190159	28963	10195160	10216996
2021-02-23	8324	8316	3689534	3689127	28939	28942	28261595	28254026	10257875	10242316
2021-02-24	8332	8335	3721061	3707712	28947	28944	28336097	28326517	10324463	10302692
2021-02-25	8349	8343	3746475	3740348	28957	28951	28413388	28405129	10390461	10370618
2021-02-26	8358	8363	3746707	3767587	28965	28963	28486394	28486268	10455630	10438595

Table 4. Real values vs. Values predicted by Holt with Dampened trend

Date	Mali		France		Australia		USA		Brazil	
	Real values	Predicted values	Real values	Predicted values	Real values	Predicted values	Real values	Predicted values	Real values	Predicted
2021-02-22	8306	8310	3669354	3684567	28937	28934	28190159	28204446	10195160	10214138
2021-02-23	8324	8316	3689534	3688084	28939	28941	28261595	28252513	10257875	10237408
2021-02-24	8332	8335	3721061	3706683	28947	28944	28336097	28325049	10324463	10299422
2021-02-25	8349	8343	3746475	3739307	28957	28951	28413388	28403632	10390461	10369579
2021-02-26	8358	8360	3746707	3766522	28965	28963	28486394	28484712	10455630	10439238

Table 5. Real values vs. Values predicted by Holt with exponential trend

Date	Mali		France		Australia		USA		Brazil	
	Real values	Predicted values	Real values	Predicted values	Real values	Predicted values	Real values	Predicted values	Real values	Predicted
2021-02-22	8306	8306	3669354	3686297	28937	28934	28190159	28206524	10195160	10217941
2021-02-23	8324	8324	3689534	3689712	28939	28942	28261595	28254376	10257875	10241040
2021-02-24	8332	8332	3721061	3708277	28947	28943	28336097	28326863	10324463	10303092
2021-02-25	8349	8349	3746475	3740981	28957	28952	28413388	28405503	10390461	10373401
2021-02-26	8358	8358	3746707	3768269	28965	28964	28486394	28486679	10455630	10443249

6 Conclusion

In this paper, we have studied the problem of predicting the daily total number of reported COVID-19 cases in a given country. As a solution, we have proposed a Holt smoothing based prediction model combined with a logarithm function for cold start purposes. Our intensive experimental evaluation conducted on various datasets has showed that the proposed model with a linear trend presents very good performances. We have also proved the robustness and an easy generalization of our model to any country in the world. As perspectives, we plan to evaluate our model by using other performance metrics, to perform a deeper comparison with existing prediction models such as SIR and to improve our model so that it can do long term prediction.

References

1. Aragon, Y.: Lissage exponentiel. In: Séries temporelles avec R. Pratique R, pp. 121–132. Springer, Paris (2011). https://doi.org/10.1007/978-2-8178-0208-4_6
2. Chimmula, V.K.R., Zhang, L.: Time series forecasting of COVID-19 transmission in Canada using LSTM networks. Chaos Solitons Fractals **135**, 109864 (2020)
3. Dufour, J.M.: Lissage exponentiel. Université de Montréal (2002)
4. Indhuja, M., Sindhuja, P.: Prediction of COVID-19 cases in India using prophet. Int. J. Stat. Appl. Math. **5**, 4 (2020)
5. Karako, K., Song, P., Chen, Y., Tang, W.: Analysis of COVID-19 infection spread in Japan based on stochastic transition model. Bioscience trends (2020)
6. Ndiaye, B.M., Tendeng, L., Seck, D.: Analysis of the COVID-19 pandemic by SIR model and machine learning technics for forecasting. arXiv preprint arXiv:2004.01574 (2020)
7. World Health Organization: 2019 World COVID-19 Report, December 2019. https://www.who.int/emergencies/diseases/novel-coronavirus-2019
8. Zareie, B., Roshani, A., Mansournia, M.A., Rasouli, M.A., Moradi, G.: A model for COVID-19 prediction in Iran based on China parameters. MedRxiv (2020)
9. Zhao, S., et al.: Preliminary estimation of the basic reproduction number of novel coronavirus ((2019-nco)v) in China from 2019 to 2020: a data-driven analysis in the early phase of the outbreak. Int. J. Infect. Dis. **92**, 214–217 (2020)

Learning Color Transitions to Extract Senegalese License Plates

Modou Gueye[(✉)] [iD]

Université Cheikh Anta Diop, Dakar, Senegal
modou2.gueye@ucad.edu.sn

Abstract. Automatic License Plate Recognition (ALPR) is gaining increasing attention. Indeed it offers a wide range of applications such as automatic toll, parking payment and the control of the compliance with traffic laws.

The detection and extraction of the license plate is the most important step of an ALPR task as it determines the speed and robustness of the system. In this work, we deal with the issue of detecting and extracting senegalese license plates from car images.

Our method is based on the learning of the color transitions specific to the license plate. Then it uses vectors of features to describe textures of the different regions of the image. From these vectors, an analysis of the probability distribution allows to find the license plate region and to extract it.

Our experimentation on a dataset of 182 car images shows the effectiveness of our method which achieves a detection accuracy around 92%.

Keywords: License plate extraction · Artificial neural networks · ALPR · Color transition

1 Introduction

Government policies are increasingly oriented towards intelligent transport systems (ITS) for a better management of road traffic in major urban centers. These systems enable safer, better coordinated and "smarter" use of transport networks through the use of traffic information. Among the most popular solutions, Automatic License Plate Recognition (ALPR) is gaining more and more attention [11,14].

While making it possible to strengthen compliance with traffic laws with the use of surveillance cameras, ALPR also makes it possible to automate traffic control and payment for services such as tolls and parking [8]. Thus it offers a wide range of applications.

However, ALPR is usually computationally intensive, especially when the input image needs to be processed as a whole. In practice, license plate recognition is done in four steps as illustrated in Fig. 1.

Published by Springer Nature Switzerland AG 2021. All Rights Reserved
Y. Faye et al. (Eds.): CNRIA 2021, LNICST 400, pp. 28–39, 2021.
https://doi.org/10.1007/978-3-030-90556-9_3

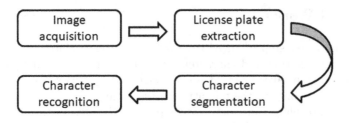

Fig. 1. ALPR processing steps

First, an input image from a video sequence is given to the system. The latter processes the image in its entirety in order to detect the position of the plate and extract it. Once the plate is extracted, it is segmented into characters and then an optical character recognition (OCR) algorithm is then performed for each segmented character to determine the corresponding letter or number.

The detection and extraction of the license plate region is the most important step, nay the most difficult, as it determines the speed and robustness of the system. Several studies and researches have been devoted to this step. Thus different methods have been proposed like edge detection, morphological operations and features selection such as color [13].

In this paper, we are interested in the detection and extraction of senegalese license plates. Indeed, senegalese government has been working for more than a year to standardize senegalese license plates. But to our knowledge, there is currently no solution for automatic recognition of new senegalese license plates, although this could greatly improve the urban mobility by strengthening compliance with traffic laws.

We aim to build a "lightweight" ALPR adapted to operate under limited capacity devices by consuming less power while operating quickly and accurately. Our proposal relies on two neural networks to learn the specific color transitions inside new senegalese license plates. With this capability, we are able to handle input images of various scenes and under different lighting conditions with sufficient training data. The experiments we carried out on a dataset of 182 images containing cars with senegalese license plates demonstrate the effectiveness of our technique.

The remainder of this article is organized as follows. In Sect. 2, we present works related to automatic detection of license plates in an image, in particular those based on the analysis of edges and the features of the license plate region such as its foreground and background colors. In Sect. 3 we detail our proposed license plate detection method based on the analysis of color transitions. We expose its functioning and explain how it proceeds to locate a vehicle license plate and extract it. Then in Sect. 4 we show our experimental results. Finally, in Sect. 5 we conclude this paper and present some future works.

2   Related Works

The most important step of an ALPR task is the license plate extraction as it allows the segmentation and character recognition steps to be done correctly. The extraction step takes as input a car image and returns as output a region of the image that contains the potential license plate.

As the license plate can exist anywhere in the image, all the pixels of the image have to be considered. But processing every pixel in the image may lead to an expensive processing time. Instead, most ALPR systems rely to the custom license plate features derived from its format and characters. Therefore they can process only the pixels that have these features.

The colors of the license plate are one of the used features since the colors of license plates are legislated in many countries. Likewise, the rectangular shape of the license plate boundary is another feature that can be considered. The license plate texture known as the color change between the characters foreground and the license plate background can also be used. Finally, several features can be combined to locate the license plate.

In the following sub-sections, we present a summary of some methods used to detect license plates in order to extract it.

2.1   License Plate Extraction Using Edge Detection

Considering that every license plate is rectangular and has a known aspect ratio, it can be extracted by finding all possible rectangles in the image.

Edge detection methods are commonly used to find the rectangles. First, Sobel filter is applied to the input image for edge detection, then Hough Transformation is applied to identify the straight lines in the image.

Once the edges retrieved and some morphological steps that eliminate unwanted edges done, the license plate rectangle is located by using geometric techniques for detecting edges that form a rectangle. Then the rectangles that have the same aspect ratio than the license plate are considered as candidates.

Edge detection methods are simple and fast with a good extraction rate under various illumination conditions. However, they need the continuity of the edges and they are highly sensitive to unwanted edges. Therefore they are not suitable to be used with blurry and complex images [1,4,13].

2.2   License Plate Extraction Using Texture Features

Many countries have legislated the colors of their license plates with very contrasting colors. Texture-based methods are based on the hypothesis that the presence of characters in the license plate should result in a significant color difference between the plate and its characters. Hence, the license plates are viewed as irregularities in the texture of the image. Therefore, the abrupt changes in the local features are the potential license plate. Indeed, the color transition makes the plate region to have a high edge density.

All the texture-based methods are robust against license plate deformation and it is a key advantage of using these methods. Still, these methods involve complex computations and work poorly with complex backgrounds and different illumination conditions [1, 4, 13].

2.3 License Plate Extraction Using Color Features

Color-based methods rely on the fact that the color of a license plate is different from the background color of the car. More precisely, the color combination of the plate and its characters is unique and is nowhere found in the image other than in the plate region [5].

Some color-based methods classify all the pixels in the input image using HSB or RGB color model. They commonly use neural networks to classify the color of each pixel [13]. Then the highest color density region is taken as the license plate region. While some others focus on color transitions, they are referenced as color edge detectors. In contrary to the first methods, the latter offer more robustness as they can handle input images from various scenes and under different conditions [5].

Extracting license plate using color information has the advantage of detecting inclined and deformed plates but it is sensitive to illumination changes.

Besides, it makes wrong detections especially when some another part of the image have similar color than the license plate. To overcome this issue, statistical threshold can be adopted to select candidate regions.

Color-based methods are often combined with some other technique to achieve accurate results [4, 13].

2.4 License Plate Extraction Using Character Features

License plate extraction methods based on locating its characters have also been proposed. These methods consider all regions with characters as possible license plate regions. They are robust and can achieve good results under different illumination conditions and viewpoints.

Due to the fact that they scan the whole image for the presence of characters, they are time-consuming and often prone to errors if there are other texts in the image [4, 13].

2.5 Discussion

In order to effectively detect the license plate, many methods combine two or more features of the license plate. They are called hybrid extraction methods. Color feature and texture feature are frequently combined. Our proposal combines these two features in order to take advantage of their strengths while avoiding their weakness.

According to the recent development in computer vision approaches, most of the statistical methods have been replaced by deep learning neural networks due to their high accuracy in object detection. Embracing this fact, many studies in license plate detection have used different types of neural networks [3,13].

Several studies have used the state-of-the-art YOLO object detector for license plate detection [3,13]. They often use two separate CNNs for vehicle detection and license plate detection.

The accuracy of deep learning methods is slightly better than statistical methods, but they fail on the computational time aspect to be used in a real-time context with limited capacity devices. This explains why we prefer to use a multiple-ordinate neural network architecture rather than a deep one.

A complete review of all these techniques can be found in these surveys [1,2,6,13].

3   Learning Color Transitions to Extract Senegalese License Plates

The senegalese authorities have chosen to overhaul the vehicle registration system to fight fraud. Thus the new license plates are on a white background with black writing. A vertical blue stripe is positioned at the start of the plate and the numbering is done in seven alphanumeric positions. Figure 2 shows an example of new senegalese license plates.

Fig. 2. Senegalese license plate

The main idea of our proposal relies on the detection of some color transitions specific to senegalese license plates. Our approach uses a multiple ordinate neural network architecture (MONNA) made of two small neural networks followed by a recomposition unit which classifies each pixel of the image whether it belongs to the license plate region or not. Each of the two neural networks is a binary classifier implemented by a Multi-Layer Perceptrons (MLP) with a single hidden layer. Their outputs serve as entries of the recomposition unit which processes them to determine the class of each pixel of the image.

This way of doing things is not new [7]. Indeed the use of individual classifiers and simpler feature extractions allow to create relatively shallower models that allow significantly faster processing compared to some Deep learning techniques. Though this kind of model is relatively simple, they are currently the most accurate models for license plate detection [9,13].

Each of the two binary classifiers receives as input a sequence of five pixels that is to be classified as belonging or not to the license plate region.

Following this classification, each row of the image is represented by a statistical descriptor making it possible to calculate the probability that it covers the license plate region. From there, the position of the license plate can be easily determined by the distribution of the probabilities and then the license plate can be extracted. Let us notice here that we introduce a new probabality measure which is one of our main contributions as we will describe below.

In theory, all the pixels of the input image should be processed. However, to greatly reduce the cost of traveling through all of them, a vertical edges detection is applied in the preprocessing step to determine the pixels of interest (POI) to be processed. Only the POIs are processed as they indicate color transitions into the image. Each of POI is associated with the four pixels of the image which follow it in order to constitute a sequence of five pixels which is transmitted to the classifiers. These latter determine whether or not the sequence contains a color transition specific to the presence of the license plate. Figure 3 sums up the main steps of our license plate extraction method.

In the following sub-sections, we detail each of these steps and its task in the processing.

3.1   Image Preprocessing

As said before, a preprocessing step is carried out on the image before its submission to the classifiers. The objective of this step is to reduce the processing time of the classification by determining in advance the regions of the image which may contain a part of the plate.

Beforehand, the image is grayscaled and its contrast enhanced by histogram equalization. This is the object of our first convolution on the image with the row vector $[0.3, 0.4, 0.3]$ as the kernel. Then a second convolution with the row vector $[1, -1]$ is applied to detect vertical edges making it possible to determine abrupt color transitions separating the foreground from the image background. Finally, Gaussian blurring followed by a dilation of the detected edge regions is done in order to prepare the sequences of five pixels on which the two binary classifiers must work. Figure 3(b) shows an overview of the obtained result at this phase.

Once completed, the resulting image is binarized with its black pixels representing the POIs to be considered. From this binary image, one can easily make the correspondence of its POIs with the pixels of the real input image. And thus find the same POIs in the input image. All the other pixels are ignored in the classification phase.

This preprocessing makes it possible to reduce the computational time of the classification phase of the pixels of the image by 85% on average.

3.2   POI Classification

In this step, the two binary classifiers work to find the regions of the image containing specific features of senegalese license plates. The use of the two classifiers allows us to break down the classification problem into two lesser complex sub-problems. The responses of the classifiers are processed by a recomposition unit.

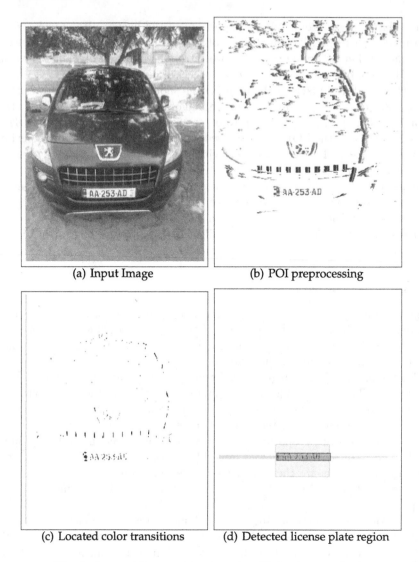

(a) Input Image

(b) POI preprocessing

(c) Located color transitions

(d) Detected license plate region

Fig. 3. Steps of license plate detection (Color figure online)

The preprocessed image is submitted to the two binary classifiers. The first classifier classifies each sequences of five pixels and determines whether or not it corresponds to *blue-white* color transitions. While the second works on *black-white* color transitions. They learn from a training set of ten car images with different scenes and car body colors.

Blue-White Color Transition Classifier. We used a MLP with a single hidden layer of twenty neurons. The input layer has five neurons representing the sequences of five pixels to be taken as input. The output layer determines whether the sequence corresponds in a transition from blue to white color.

The color of each pixel in the image is represented in the RGB color model. Each neuron in the input layer is activated according to the ratio β of blue to red and green via the following formula:

$$\beta = \frac{B + B}{1 + R + G} \tag{1}$$

Black-White Color Transition Classifier. In parallel with the first classifier, the second one searches for black-white color transitions. It is also a MLP with a single hidden layer which is made up of twenty neurons but it has ten neurons in its input layer. Indeed, the pixels are taken in the HSB color model and two input neurons describe the saturation and brightness of each of the five pixels of the input sequences. Let us remind that the HSB color model is an alternative representation of the RGB color model and it is designed to better adapt with the way human vision perceives colors. Therefore it models how colors appear under light and is more suited than RGB color model to distinguish dark pixels from light ones corresponding to black-white color transitions.

3.3   Plate Area Feature Descriptors

Once the processing of the two classifiers is complete, their outputs are combined by the recomposition unit. Figure 3(c) points out the result obtained which is a sparse matrix whose values represent blue-white or black-white color transition positions.

Following the image filtering by the recomposition unit, a statistical analysis of the filtered image is carried out on each of the image rows. Thus each row is associated with a numerical vector descriptive of the successions of black-white color transition preceded by a blue-white color transition. The numerical vector reflects the coloring features specific to senegalese license plates found on the corresponding row. As a reminder, Fig. 2 shows a senegalese license plate example.

The first entry of the vector represents the position of the blue-white color transition preceding the list of black-white color transitions. In the absence of a blue-white color transition on the row, the vector remains empty. The following entries of the vector record teh positions on the row of the sequence of black-white color transitions following the blue-white one. The maximum size of the vector is fixed to 15 entries to store the position of the blue-white color transition in front of the fourteen positions of the black-white color transitions which compose the seven characters of the license plate.

3.4   Probability Measure of License Plate Presence

An array of vector descriptors representing all the rows of the image is made. The distribution of transition positions into the vectors allows to locate the region of the license plate. Table 1 gives an example of such a distribution corresponding to the rows of the license plate region in Fig. 3(d).

Table 1. Example of the distribution of color transition positions in the plate region

129	145	156	172	177	192	207	214	220				
127	145	156	172	177	191	208	214	220				
127	145	156	172	182	192	203	208	214	220			
127	141	145	152	157	171	183	193	198	203	208	214	220
127	141	146	152	157	170	183	194	203	208	214	220	
128	141	146	152	157	170	183	194	208	214	220		
129	146	157	169	183	194	208	214	220				
130	146	157	168	177	183	188	193	202	208	214	219	
124	140	147	152	158	167	182	193	202	218			
129	140	147	151	158								

For each of the vectors, we calculate the probability that it overlaps the license plate region on the interval of positions defined by the vector. We introduce a new probability measure that calculates the entropy of the dispersion of black and white colors over the interval. Then we associate it by multiplication with the Gini index' value of the inequality of the distances between the black-white color transition positions representing the locations of the characters on the plate. The combination of the values of entropy and Gini index makes it possible to easily find the most probable region where the plate is on the input image. We achieve it with the establishment of the histogram of the two combined values.

In Fig. 3(d), we illustrate the rows of the input image that most likely overlap the license plate region. They are colored with a green background color relatively to their probability values.

A similarity analysis of the neighborhood of the largest peak allows to calculate the height in pixels of the license plate region. As the ratio of the width and the height of senegalese license plates is around 3.75, we can deduce the width of the license plate if the latter is not tilted on the image. Otherwise, we apply a geometric transformation to correct tilt license plate. Such transformations are popular and widely presented in [10, 12].

4   Experimentation

We demonstrate in this section the effectiveness of our proposal. In the previous section, we explain its functioning. In this section, we describe how we did our experimental evaluation and we present the accuracy of our method to locate senegalese license plates in a car image.

4.1 Dataset

In our knowledge, there is no senegalese license plate dataset publickly available to researchers, thus we made our own dataset which contains 182 labeled car images with the bound of the senegalese license plate within the image. We collected the images from senegalese websites specializing in the sale of cars[1].

The images in the dataset have different sizes and are taken from different scenes. Furthermore, the cars in the images have various body colors.

4.2 Experimental Results

Our evaluation consisted to train our classifiers with a training set of ten cars with different car body colors in addition to different brightness and license plate skews. Then we submit the rest of the images for the test phase.

Our system extract for each image the license plate region which is compared to the real license plate region into the image. We achieve a detection accuracy around 92%.

Figure 4 shows a sample of detected license plates among the images of our test set.

Fig. 4. License plate detection

---

[1] The dataset is available at https://bit.ly/3rd0ofR.

5 Conclusion

In this paper, we presented a novel senegalese license plate extraction algorithm. We detailed its functioning and discussed how it extract license plate from an input image. The experiments we led points out the effectiveness of its detection. Therefore license plates can be accurately extracted.

As part of future work, we aim to collect a larger senegalese license plate dataset to better improve the accuracy of our detection by further training the binary classifiers. Next, we intend to compare our approach to others that are available in the literature.

Likewise, we aim to extend our method to detecting multiple license plates in a single image.

References

1. Atiwadkar, A., Patil, K., Mahajan, S., Lande, T.: Vehicle license plate detection: a survey (2015)
2. Bhatti, M.S., Saeed, F., Ajmal, M., Tayyab, M., Naeem, Q., Safdar, A.: Survey of computer vision techniques for license plate detection (2014)
3. Bhujbal, A., Mane, D.: A survey on deep learning approaches for vehicle and number plate detection. Int. J. Sci. Technol. Res. **8**, 1378–1383 (2020)
4. Du, S., Ibrahim, M., Shehata, M.S., Badawy, W.M.: Automatic license plate recognition (ALPR): a state-of-the-art review. IEEE Trans. Circuits Syst. Video Technol. **23**(2), 311–325 (2013)
5. Lee, E.R., Kim, P.K., Kim, H.J.: Automatic recognition of a car license plate using color image processing. In: Proceedings of 1st International Conference on Image Processing, vol. 2, pp. 301–305 (1994)
6. Farajian, N., Rahimi, M.: Algorithms for licenseplate detection: a survey. In: 2014 International Congress on Technology, Communication and Knowledge (ICTCK), pp. 1–8 (2014)
7. Gratin, C., Burdin, H., Lezoray, O., Gauthier, G.: Classification par réseaux de neurones pour la reconnaissance de caractères. Application à la lecture de plaques d'immatriculation. In: 34ème Journée ISS France, 2 pages, France (2011)
8. Haripriya, K., Harshini, G., Sujihelen: Survey on efficient automated toll system for license plate recognition using open CV. In: 2018 International Conference on Emerging Trends and Innovations in Engineering and Technological Research (ICETIETR), pp. 1–6 (2018)
9. Lezoray, O., Fournier, D., Cardot, H., Revenu, M.: MONNA: a multiple ordinate neural network architecture, August 2002
10. Liang, Z., Zhang, S.: A configurable tilt license plate correction method based on parallel lines. Appl. Mech. Mater. **1429–1433**(08), 385–386 (2013)
11. Mani Raj, S.P., Rupa, B., Sravanthi, P.S., Sushma, G.K.: Smart and digitalized traffic rules monitoring system. In: 2018 3rd International Conference on Communication and Electronics Systems (ICCES), pp. 969–973 (2018)
12. Nguyen, C., Binh, N., Chung, S.-T.: Reliable detection and skew correction method of license plate for PTZ camera-based license plate recognition system, pp. 1013–1018, October 2015

13. Shashirangana, J., Padmasiri, H., Meedeniya, D., Perera, C.: Automated license plate recognition: a survey on methods and techniques. IEEE Access **9**, 11203–11225 (2021)
14. Shreyas, R., Kumar, B.V.P., Adithya, H.B., Padmaja, B., Sunil, M.P.: Dynamic traffic rule violation monitoring system using automatic number plate recognition with SMS feedback. In: 2017 2nd International Conference on Telecommunication and Networks (TEL-NET), pp. 1–5 (2017)

Intrusions Detection and Classification Using Deep Learning Approach

Léonard M. Sawadogo[✉], Didier Bassolé, Gouayon Koala, and Oumarou Sié

Laboratoire de Mathématiques et d'Informatique, Université Joseph KI-ZERBO,
Ouagadougou, Burkina Faso
http://www.univ-ouaga.bf

Abstract. In this paper we propose an intrusions detection technique using Deep Learning approach that can classify different types of attacks based on user behavior and not on attacks signatures. The Deep Learning approach used is Supervised Learning model called Convolutional Neural Networks (CNN) coupled with Tree Structure whose set is named Tree-CNN. This structure allows for incremental learning. This makes the model capable of learning how to detect and classify new types of attacks as new data arrives. The model was implemented with Tensor-Flow and trained with the CSE-CIC-IDS2018 dataset. We evaluated the performance of our proposed model and we made comparisons with other approaches considered in related works. The experimental results show that the model can detect and classify intrusions with a score of 99.94% for the detection and 97.54% for the classification.

Keywords: Intrusion detection · Deep learning · Classification · Tree-CNN

1 Introduction

Intrusions generally involve gaining unauthorized access to data on a computer system or network by bypassing or defusing the security devices in place. Intrusion detection systems for the security of computer systems are diversified, but they are still confronted with two major problems, namely the rate of false alarms and the capacity to detect new attacks, in particular "zero-day" attacks. The goal of intrusion detection is to spot the actions of an attacker attempting to take advantage of system vulnerabilities to undermine security objectives. Different categories of intrusion detection methods are explored in the literature, in particular those based on a behavioral approach such as static analysis, Bayesian analysis, neural networks and those based on a scenario approach such as the search for signatures, pattern matching, simulation of Petri networks.

An intrusion detection system can attempt to identify attacks by relying on information relating to transitions taking place in the system (execution of certain programs, certain sequences of instructions, arrival of certain packets

© ICST Institute for Computer Sciences, Social Informatics and Telecommunications Engineering 2021
Published by Springer Nature Switzerland AG 2021. All Rights Reserved
Y. Faye et al. (Eds.): CNRIA 2021, LNICST 400, pp. 40–51, 2021.
https://doi.org/10.1007/978-3-030-90556-9_4

network, ...) or by studying state of certain properties of the system (integrity of programs or stored data, user privileges, rights transfers, ...). How to detect different network attacks, especially ones that have never been seen before, is a key question researchers are trying to solve. To this end, to be able to assess the possible limits of current existing intrusion detection systems and to consider the contribution of Machine Learning techniques to improve intrusion detection systems, their intrusion detection process and also efficiently manage alerts, is a major concern in the field of computer security.

The rest of this paper is structured as follows: Sect. 2 deals with related works. Section 3 presents our Approach and architecture related vulnerabilities and the classification of machine learning techniques used to predict the type of vulnerability. Section 4 illustrates the description of our dataset used, the experimental set-up and our learning algorithm. Section 5 provide discussions on results of our analysis and make comparison with other approaches. We conclude this work in the Sect. 6.

2   Related Works

Intrusion Detection System (IDS) research, particularly that involving machine learning methods, is of great interest to computer security researchers. These authors deal with various machine learning methods (supervised, unsupervised, etc.). As data issues are very important for machine learning, several researchers deal with them. These authors discuss the quantity, quality, availability and compatibility with different machine learning methods.

In [1], Alex Shenfield et al. present an approach to detect malicious attacks traffic on networks using artificial neural networks based on deep packet inspection of the network's packets. The KDD CUP 1999 dataset was used for training. They achieved an average accuracy of 98% and an average false positive rate of less than 2%. This shows that the proposed classification technique is quite robust, accurate and precise, but the false alarm rate is still high.

N. Chockwanich et al. [2] used Recurrent Neural Network (RNN) and Convolutional Neural Network (CNN) to classify five types of attack using Keras with TensorFlow. To evaluate performance, they used the MAWI dataset which are pcap ("packet capture") files and compared their results with those of Snort. The results show that Snort could not detect the network scan attack via ICMP and UDP. They proved that RNN and CNN can be used to classify Port scan, Network scan via ICMP, Network scan via UDP, Network scan via TCP, and DoS attack with high accuracy. RNN offers an accuracy of 99.76% and 99.56% for CNN.

In [3] Leila Mohammadpour et al. used CNN and the NSL-KDD dataset for network intrusion detection system. They obtained, for their experimental results, a detection rate of 99.79%.

Peng Lin et al. [4] use Long-Short Term Memory (LSTM) to build a deep neural network model and add an attention mechanism (AM) to improve the model performance. They use the CSE-CIC-IDS2018 data set. Experimental results show an accuracy of 96.2%, and they claim that this figure is better than the figures of other machine learning algorithms. Nevertheless the score is below that of other models using the same data set as shown in the work below.

In [5] V. Kanimozhi et al. compared the performance of several algorithms on the CSE-CIC-IDS2018 dataset. They showed that the artificial neural networks (ANN) performed significantly better than the other models with a score of 99.97% and an accuracy of 99.96%.

V. Kanimozhi et al. [6] propose a system which consists in detecting a botnet attack classification. The proposed system is created by applying an artificial neural network model to a CSE-CIC-IDS2018 data set. The proposed system offers a precision score of 99.97% and the average false positive rate is only 0.1%. However, the model is only applied to botnet intrusions and therefore does not have the capability to detect other types of attacks.

3   Methodology: Approach and Architecture

3.1   Approach

In order for Intrusion Detection Systems to perform well with an excellent false alarm rate and "zero-day" attack detection, they must be trained regularly with new data obtained from monitoring network traffic. More the IDS model is trained, more it will adjust to improve. Traditionally, training for artificial intelligence models takes place in one go ("One-Shoot") and then these models are used, whatever the duration, without learning new information. Different techniques including continuous learning and incremental learning [7,8] are introduced in an attempt to learn continuously. This has become a very fashionable field of research [7,9–11]. However, changing part of the parameter space immediately affects the model as a whole [12]. Another problem related to the progressive training of a Deep CNN model is the issue of catastrophic oblivion [13]. When a formed Deep CNN is exclusively recycled on new data, it results in the destruction of existing features learned from previous data.

But based on an algorithm proposed by Deboleena Roy et al. [14] we can exploit the advantages of incremental learning. In their work, to avoid the problem of catastrophic forgetting, and to keep the functionalities learned in the previous task, Deboleena Roy et al. propose a network composed of CNN that develops hierarchically as that new classes are being introduced. CNN are able to extract high-level features that best represent the abstract form of low-level features of network traffic connections. We chose to use CNNs because of the many advantages they offer, particularly their ability to select the most significant features themselves and their architecture gives them the ability to prioritize the selected features from the simplest to the most sophisticated.

3.2   Architecture

Inspired by hierarchical classifiers, the model we propose, Tree-CNN is composed of several nodes linked together to have a tree structure. The first node of the tree structure is the "root" node where the first classification takes place. All the other nodes of the tree structure, except the "leaf" nodes, have a Deep Convolutional Neuronal Network (Deep CNN) which is trained to classify the entry for their "child" nodes. The classification process starts at the "root" node and the data is then passed on to the next "child" node, depending on the result of the classification of the "root" node. This node then classifies the data at its own level and transmits it in turn, depending of the result of the classification, to one of its "child" nodes. This process is repeated until a "leaf" node is reached. This is the end of the classification steps.

The Fig. 1 shows a three-level architecture of our model. The method of training an artificial neural network with such an architecture is described by the Algorithm 1. One of the advantages of the tree structure is that it allows a considerable reduction of the decision (prediction) time. The principle is to first detect whether or not it is an attack and then to classify the intrusion. This is done by a structure in the form of a tree as shown in Fig. 2. There are three levels:

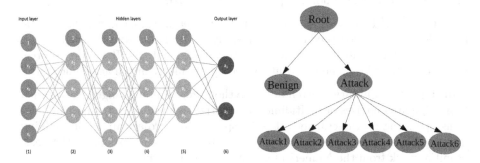

Fig. 1. Structure of a CNN   **Fig. 2.** Illustration of tree hierarchie

- the first level is the "root" node;
- the second level is made up of two nodes: a "leaf" node linked to the Benign class and a node linked to the Attack class;
- the third level contains "leaf" nodes, all from the "Attack" node.

All nodes, except the "leaf" nodes, contain a Deep CNN which is trained to classify the classes of its "child" nodes, each at its own level. The first node which is the root node is used to make the decision (to predict) whether it is an attack or a benign event. And if it's an attack, to give more details about the attack so that the administrator can intervene effectively, the node linked to the "Attack" class is led to make a classification. The node linked to the Attack class classifies the different types of attacks. The fourteen types of attacks, grouped into seven, are classified by this node.

4 Implementation

4.1 Tools Used

TensorFlow.[1] It is an open-source Google platform dedicated to machine learning. It offers a complete and flexible ecosystem of tools, libraries and of community resources to enable researchers to advance in the machine learning domain, and for developers to create and deploy applications that exploit this technology.

Pandas.[2] It is an open source library used for data manipulation and processing. It is a fast, powerful, flexible and easy to use data manipulation and analysis tool. In particular, it offers data structures and operations for manipulating numerical tables and time series.

Keras.[3] It is an API for deep neural networks and is written in Python, and now included in TensorFlow. It focuses on ergonomics, modularity and extensibility. It was born within the framework of the ONEIROS project (Open-ended Neuro-Electronic Intelligent Robot Operating System). It was originally written by François Chollet in his book Deep learning with Python.

Scikit-Learn.[4] It is a free Python library for the learning machine. It has simple and effective tools for predictive data analysis. Built on NumPy, SciPy, and matplotlib, it is accessible and reusable in various contexts. It is open source, and is developed by numerous contributors, particularly in the academic world.

NumPy.[5] It is a library written in the Python programming language, designed to manipulate matrices or multidimensional arrays as well as mathematical functions operating on these arrays. It is an open source project aimed at enabling numerical computation with Python. It has been created in 2005, on the basis of initial work from the Numerical and Numarray libraries.

4.2 Data-Set

The dataset used is CSE CIC-IDS2018[6]. It is one of the most recent, realistic and up-to-date public data sets and more complete in terms of the types of attacks they contain. The attacks that the CSE CIC-IDS2018 contains are topical and best reflect threats networks and information systems of our companies, which are currently facing nowadays. The CSE CIC-IDS2018 dataset contains 86 "features". The data contained in CSE CIC-IDS2018 are labelled with fourteen labels which are presented in Table 2. The number of lines in the data set is over sixteen million (16,000,000) and is distributed as shown in Table 1 according to the type of attack.

---

[1] https://tensorflow.org.
[2] https://pandas.pydata.org.
[3] https://keras.io.
[4] https://scikit-learn.org.
[5] https://numpy.org.
[6] http://www.unb.ca/cic/datasets/ids-2018.html.

Table 2 clearly show that there is an imbalance in the data. The label "benign" alone represents more than 80% of the data. Indeed, in everyday use, benign cases far outweigh the cases of attack. But the model risks giving more weight to the benign label compared to others during training. This problem can be alleviated by using the SMOTE (Synthetic Minority Oversampling Technique) [15] to try out to balance the data. SMOTE is an over-sampling method, it works by creating synthetic samples from the class or the minority classes instead of creating simple copies.

Table 1. Distribution of CSE CIC-IDS2018 data

Label	Number
Benign	13484708
DDOS attack-HOIC	686012
DDoS attacks-LOIC-HTTP	576191
DoS attacks-Hulk	461912
Bot	286191
FTP-BruteForce	193360
SSH-Bruteforce	187589
Infilteration	161934
DoS attacks-SlowHTTPTest	139890
DoS attacks-GoldenEye	41508
DoS attacks-Slowloris	10990
DDOS attack-LOIC-UDP	1730
Brute force-Web	611
Brute force-XSS	230
SQL injection	87

Table 2. Breakdown of CSE CIC-IDS2018 data into "Benin" and "Attack"

Label	Number	Percentage
Benign	13484708	0,8307
Attack	2748235	0,1693

4.3 Learning Algorithm

We used Deboleena Roy et al.'s algorithm [14]. To train the model to recognize a number of new classes of attacks, we provide data on these attacks at the root node. We obtain a three-dimensional matrix at the output layer: $O^{K \times M \times I}$, where
K is the number (in our case 02) of child nodes of the root node,
M is the number of new classes of attacks and
I is the number of data samples per class.
$O(k; m; i)$ indicates the output of the k^{ith} node for the i^{ith} given belonging to the m^{ith} class, where $k \in [1, K]$, $m \in [1, M]$ and $i \in [1, I]$.
$O_{avg}(k, m)$ (Eq. (1)) is the average of the outputs(of the k^{ith} node and the m^{ith} class) on I data.
$O_{avg}^{K \times M}$ is the matrix of these averages over the I data.

The *Softmax* (Eq. (2)) is calculated on each average O_{avg} (Eq. (1)) to get a matrix $L^{K \times M}$.

An ordered list S is generated from the matrix $L^{K \times M}$, having the following properties:

- The list S has M objects. Each object is uniquely linked to one of the new classes M.
- Each object $S[i]$ has the following attributes:
  1. $S[i].label$ = label of the new class.
  2. $S[i].value = [v_1, v_2, v_3]$, the 3 highest *Softmax* values of the averages (O_{avg}) of this new class, ranked in descending order $v_1 \geq v_2 \geq v_3$.
  3. $S[i].nodes = [n_1, n_2, n_3]$, the nodes corresponding respectively to the values *Softmax* v_1, v_2, v_3.
- S is ordered by descending value of $S[i].value[1]$

$$O_{avg}(k, m) = \sum_{i=1}^{I} \frac{O(k, m, i)}{I} \tag{1}$$

$$L(k, m) = \frac{e^{O_{avg}(k,m)}}{\sum_{k=1}^{K} e^{O_{avg}(k,m)}} \tag{2}$$

This scheduling is done to ensure that new classes with high Softmax values are first added to the Tree-CNN tree. After building S, we examine its first element and take one of the three paths:

i. add the new class to an existing child node: If v_1 is greater than the next value (v_2) of a threshold, α (in our case $\alpha = 0, 1$), this indicates a strong resemblance to the node n_1. The new class is then added this node;

ii. merge 2 child nodes and add the new class: If two of the values *Softmax* are close, i.e., when $v_1 - v_2 < \alpha$, and $v_2 - v_3 > \beta$ (a user-defined threshold, here we have defined it $\alpha = 0, 1$), then, if n_2 is a leaf node, we merge n_2 into n_1 and add the new class to n_1;

iii. add the new class as a new child node: If the three values *Softmax* are not different with wide margin ($v_1 - v_2 < \alpha, v_2 - v_3 < \beta$: for example if the three largest values *Softmax* are $v_1 = 0, 35$, $v_2 = 0, 33$, and $v_3 = 0, 31$), or if all child nodes are full, the network adding the new class as a new child node.

To prevent the Tree-CNN tree from becoming unbalanced, the maximum number of children that a branch node can have can be set. The procedure described above is repeated iteratively until all new classes are assigned a location below the root node. The pseudo code is described in the paper [14]. We also illustrate an example of incremental learning in Tree-CNN with the Fig 2.

5   Tests and Results

5.1   Execution Environment

The tests were carried out in a Cloud environment with shared resources. Indeed we used Google Colab in which we have access to 25.5 Gb of RAM memory, a GPU (Tesla P100-PCIE-16 GB) and a CPU (Intel(R) Xeon(R) CPU @ 2.30 GHz). But Google does not give any guarantee on the total availability of the promised resources. Our model is composed of ten (10) layers of convolutions interspersed with two (02) layers of Maxpooling, one Flatten layer, two (02) dense layers. All the convolution and dense layers, except the last one, have a *ReLU* activation. The last layer has *Sofmax* activation.

The training has been carried out over 2000 epochs. The Algorithm 1 shows how is the inference of our model.

5.2   Score and Other Measures

Four (04) basic elements are used to assess performance IDS. They can be represented in the form of a cross table Table 3. The actual metrics are:

The metrics derived from these basic elements are:

Algorithm 1: Inference algorithm [14]

```
 1: I = Input Image, node = Root Node of the Tree
 2: procedure ClassPredict(I, node)
 3:     count = # of children of node
 4:     if count = 0 then
 5:         label = class label of the node
 6:         return label
 7:     else
 8:         nextNode = EvaluateNode(I, node)
 9:         ▸ returns the address of the child node of highest
    output neuron
10:         return ClassPredict(I, nextNode)
11:     end if
12: end procedure
```

Table 3. Basic elements for measuring the performance of an IDS

	Positive prediction	Negative prediction
Intrusions	TP (True Positive): intrusions are identified as intrusions	FN (False Negative): intrusions are identified as benigns
Benigns	FP (False Positive): benigns are identified as intrusions	TN (True Negative): benigns are identified as benigns

- **Accuracy**: it is defined as the ratio of correctly predicted samples to the total number of samples. The score is an appropriate measure only when the data set is balanced.

$$\textbf{Accuracy:} = \frac{TP+TN}{TP+FN+FP+TN}$$

- **Precision**: it is defined as the ratio of correctly predicted positive samples to predicted positive samples. It represents confidence in the detection of attacks.

$$\textbf{Precision:} = \frac{TP}{TP+FP}$$

- **Recall**: it is defined as the ratio of correctly predicted positive samples to the total number of actually positive samples. It reflects the system's ability to recognize attacks.

$$\textbf{Recall:} = \frac{TP}{TP+FN}$$

- **F1-measure**: It is defined as the harmonic mean of Precision and Recall. The higher rate of F1-measure shows that the system is performed better

$$\textbf{F1-measure:} = 2 \times \left[\frac{Precision \times Recall}{Precision + Recall} \right] = \frac{2 \times TP^2}{TP^2 + TP \times (FN+FP)}$$

- **Rate of False Negative (RFN)**: it is defined as the ratio of intrusions are identified as benigns to the total number of intrusions. The RFN is also termed the Missed Alarm Rate.

$$RFN = \frac{FN}{TP+FN}$$

- **Rate of False Positive (RFP)**: it is defined as the ratio of benigns are identified as intrusions to the total number of benign. The RFP is also termed the False Alarm Rate.

$$TFP = \frac{FP}{VN+FP}$$

The tests for detection (performed by the "root" node) give the following results Table 4:

Table 4. Results of detection test

Accuracy	Precision	F1-measure	Recall	RFP	RFN
99,94%	99%	99%	99%	0.0001	0.0001

For the multi-class classification (performed by the "Attack" node), we obtain a accuracy of *97.54%*. The following Table 5 gives details.

Table 5. Results classification test

Classe	Precision	F1-measure	Recall
DDOS	100%	100%	100%
DoS	96,50%	92,71%	89,21%
Bot	100%	99,82%	99,64%
BruteForce	90,00%	93,43%	97,12%
Infilteration	97,19%	98,40%	99,64%
Web attack and injection	99,63%	98,73%	97,84%

5.3 Comparison with Related Works

It is difficult to compare work that does not use the same methods or dataset. So we have chosen to compare with work that uses similar machine learning techniques to our own. We have through the Table 6 summarized the results obtained in other works, particularly those using deep learning. In the cited works, some of them deal with intrusion classification (multi-classes) Table 7 and others deal with intrusion detection Table 6 (binary classification).

As far as detection is concerned, we obtained a score of 99.94%, an accuracy of 99%. These measurements are slightly lower than those of Kanimozhi et al. [6] (the best) cited in Table 6. But this can be explained by the fact that our model takes more parameters into account since, in addition to detection, it makes the classification, contrary to that of Kanimozhi et al. Also through the Table 6, we see very clearly that our model is better for multi-class classification. Indeed, the score largely exceeds those of the multi-class classification works cited in Table 6. The accuracy of our model on each class shows that it is not very wrong on classification. From all the above, we can say that our method is a good means of detecting and classifying intrusions.

Table 6. Table of detection performance of some models

Works	Methods	Dataset	Accuracy	Precision	F1-measure
T. Le et al. [16]	LSTM:binary	KDD Cup99	97.54	0.97	–
Zeng et al. [17]	1D-CNN:binary	ISCX2012	99.85	–	–
Zeng et al. [17]	LSTM:binary	ISCX2012	99.41	–	–
Kanimozhi et al. [6]	ANN:binary	CSE-CIC-IDS 2018	99,97	1,0	1,0
Our method	**Tree-CNN:binary**	**CSE-CIC-IDS 2018**	**99,94**	**0.99**	**0.99**

Table 7. Table of classification performance of some models

Works	Methods	Dataset	Accuracy	Precision	F1-measure
Q. Niyaz et al. [18]	RNN:5-class	NSL-KDD	79.10	–	–
S. Potluri et al. [19]	CNN:5-class	NSL-KDD	91.14	–	–
D. Yalei et al. [20]	CNN:5-class	NSL-KDD	80,13	–	–
S. Potluri et al. [21]	DBN+SVM: 5-class	NSL-KDD	92,06	–	–
Our method	**Tree-CNN: 6-class**	**CSE-CIC-IDS 2018**	**97,53**		

6   Conclusion

We proposed in this paper a model of Intrusion Detection System using CSE-CIC-IDS-2018 and Tree-CNN, a hierarchical Deep Convolutional Neural Network for incremental learning. We evaluated the performance of the model by comparing it with other approaches. Through this evaluation, we found that our model is more effective on multi-class classification than the others mentioned in the related works. Our approach therefore allows the classification and detection of intrusions with good performance. Nevertheless, there is still room for improvement. Research carried out in this paper can be further investigated in order to improve our approach and methods used. In terms of perspectives, this means:

- improve the model: there is still room for improvement in performance, for the multi-class classification of the model by further adjusting the hyper-parameters. In addition, the model can be trained with other recent datasets in order to reduce its false alarm rates and make it even more robust and reliable;
- from IDS to IPS: another way to improve the proposed solution would be to combine our model with rule-based intrusion detection tools such as Snort, which could speed up the detection of trivial or recurring cases. In addition, it would be interesting to further develop our solution so that it can trigger actions based on detected intrusions, i.e. it can act against an attack while waiting for the administrator's intervention. The final solution would no longer be an Intrusion Detection System (IDS), but an Intrusion Prevention System (IPS).

References

1. Shenfield, A., Day, D., Ayesh, A.: Intelligent intrusion detection systems using artificial neural networks. ICT Express **4**(2), 95–99 (2018). SI on Artificial Intelligence and Machine Learning
2. Chockwanich, N., Visoottiviseth, V.: Intrusion detection by deep learning with TensorFlow. In: 2019 21st International Conference on Advanced Communication Technology (ICACT), pp. 654–659 (2019)
3. Mohammadpour, C.S.L.L., Ling, T.C., Chong, C.Y.: A convolutional neural network for network intrusion detection system. In: Asia-Pacific Advanced Network (APAN) (2018)
4. Lin, P., Ye, K., Xu, C.-Z.: Dynamic network anomaly detection system by using deep learning techniques. In: Da Silva, D., Wang, Q., Zhang, L.-J. (eds.) CLOUD 2019. LNCS, vol. 11513, pp. 161–176. Springer, Cham (2019). https://doi.org/10.1007/978-3-030-23502-4_12
5. Kanimozhi, V., Prem Jacob, T.: Calibration of various optimized machine learning classifiers in network intrusion detection system on the realistic cyber dataset CSE-CIC-IDS 2018 using cloud computing. Int. J. Eng. Appl. Sci. Technol. **4**, 2455–2143 (2019)

6. Kanimozhi, V., Jacob, T.P.: Artificial intelligence based network intrusion detection with hyper-parameter optimization tuning on the realistic cyber dataset CSE-CIC-IDS 2018 using cloud computing. In: 2019 International Conference on Communication and Signal Processing (ICCSP), pp. 0033–0036 (2019)

7. Giraud-Carrier, C.: A note on the utility of incremental learning. AI Commun. **13**(4), 215–223 (2000)

8. Ring, M.B.: Child: a first step towards continual learning. In: Thrun, S., Pratt, L. (eds.) Learning to Learn, pp. 261–292. Springer, Boston (1998). https://doi.org/10.1007/978-1-4615-5529-2_11

9. Polikar, R., Upda, L., Upda, S.S., Honavar, V.: Learn++: an incremental learning algorithm for supervised neural networks. IEEE Trans. Syst. Man Cybern. Part C (Appl. Rev.) **31**(4), 497–508 (2001)

10. Shin, H., Lee, J.K., Kim, J., Kim, J.: Continual learning with deep generative replay. In: Advances in Neural Information Processing Systems, pp. 2990–2999 (2017)

11. Zenke, F., Poole, B., Ganguli, S.: Continual learning through synaptic intelligence. Proc. Mach. Learn. Res. **70**, 3987 (2017)

12. XIao, T., Zhang, J., Yang, K., et al.: Error-driven incremental learning in deep convolutional neural network for large-scale image classification. In: Proceedings of the 22nd ACM International Conference on Multimedia, pp. 177–186 (2014)

13. Goodfellow, I.J., Mirza, M., Xiao, D., et al.: An empirical investigation of catastrophic forgetting in gradient-based neural networks. arXiv preprint arXiv:1312.6211 (2013)

14. Roy, D., Panda, P., Roy, K.: Tree-CNN: a hierarchical deep convolutional neural network for incremental learning. Neural Netw. **121**, 148–160 (2020)

15. Chawla, N.V., Bowyer, K.W., Hall, L.O., Kegelmeyer, W.P.: Smote: synthetic minority over-sampling technique. J. Artif. Intell. Res. **16**, 321–357 (2002)

16. Le, T., Kim, J., Kim, H.: An effective intrusion detection classifier using long short-term memory with gradient descent optimization. In: 2017 International Conference on Platform Technology and Service (PlatCon), pp. 1–6 (2017)

17. Zeng, Y., Gu, H., Wei, W., Guo, Y.: Deep-full-range: a deep learning based network encrypted traffic classification and intrusion detection framework. IEEE Access **7**, 45182–45190 (2019)

18. Javaid, A., Niyaz, Q., Sun, W., Alam, M.: A deep learning approach for network intrusion detection system. In: Proceedings of the 9th EAI International Conference on Bio-inspired Information and Communications Technologies (Formerly BIONETICS), pp. 21–26 (2016)

19. Potluri, S., Ahmed, S., Diedrich, C.: Convolutional neural networks for multi-class intrusion detection system. In: Groza, A., Prasath, R. (eds.) MIKE 2018. LNCS (LNAI), vol. 11308, pp. 225–238. Springer, Cham (2018). https://doi.org/10.1007/978-3-030-05918-7_20

20. Ding, Y., Zhai, Y.: Intrusion detection system for NSL-KDD dataset using convolutional neural networks. In: Proceedings of the 2018 2nd International Conference on Computer Science and Artificial Intelligence, pp. 81–85 (2018)

21. Potluri, S., Henry, N.F., Diedrich, C.: Evaluation of hybrid deep learning techniques for ensuring security in networked control systems. In: 2017 22nd IEEE International Conference on Emerging Technologies and Factory Automation (ETFA), pp. 1–8 (2017)

Comparison Study of Short Vowels' Formant: MSA, Arabic Dialects and Non-native Arabic

Ghania Droua-Hamdani[✉]

Centre for Scientific and Technical Research on Arabic Language Development (CRSTDLA), Algiers, Algeria
gh.droua@post.com, g.droua@crstdla.dz

Abstract. The paper deals with the formants of three short vowels in the Modern Standard Arabic (MSA) language produced by 29 native and non-natives speakers. The studied vowels are /a/, /u/ and /i/. F1, F2 and F3 formants were computed from 145 sentences produced by both groups of speakers. Two experiments were conducted. The first investigation compared formants values of natives speakers with those of non-native speakers. When comparing the MSA L1 vowels against their MSA L2 counterparts, results showed a variation in vowel quality especially for the vowel /a/. A comparative analysis of formants values was also carried out with results of some eight Arabic dialect studies. The outcomes depicted variation especially in the pronunciation of the vowel /u/.

Keywords: Formants · Modern standard Arabic · Native speakers · Non-native speakers · Short vowels · Arabic dialects

1 Introduction

The speech of non-native speakers may exhibit pronunciation characteristics that result from their imperfectly learning the L2 sound system, either by transferring the phonological rules from the first language to the second language or through implementing strategies similar to those used in primary language acquisition [1]. In speech recognizing tasks, distinguishing between native speakers from non-native speakers remains a big challenge to solve. To improve speech recognizers' accuracies, a huge amount of speech data is required in the training and testing phases. Features used for this purpose are usually: MFCC, formants, etc. [2–6]. Formants, whether they are defined as acoustic resonances of the vocal tract, or as local maxima in the speech spectrum, are determined by their frequency and by their spectral width. They were widely studied in speech processing as well as in speech perception, sound production comparison within languages, second language (L2) acquisition, speech pathology studies, recognition tasks, etc. [4, 8–14

Like many other languages, the Arabic language has known several research studies on formants. Regarding the auditory level, we can cite works of [15–17] conducted either on Modern Standard Arabic or on Arabic dialects. Abou Haidar investigated the

Y. Faye et al. (Eds.): CNRIA 2021, LNICST 400, pp. 52–58, 2021.
https://doi.org/10.1007/978-3-030-90556-9_5

MSA vowels system using a set of monosyllabic words to show the vowel cross-dialectal structural stability at the perception level [15]. The comparative study was done on the vowels of eight informants from different linguistic backgrounds (Qatar, Lebanon, Saudi Arabia, Tunisia, Syria, Sudan, United Arab Emirates, and Jordan). Alghamdi based his work on six isolated Arabic syllables that were uttered by 15 native informants representing three Arabic dialects (Egyptian, Saudi and Sudanese) [16]. Newman realized an experimental study on MSA vowels in connected speech. The speech variety that was studied was the Holy Quran recitation [17].

MSA phonetic system is endowed by six vowels: three short vowels (/a/, /u/ and /i/) vs. three long vowels (/a:/, /u:/ and /i:/). However, the English language has fifteen vowel sounds. The investigation examines only short vowels extracted from the speech material. The present study examines vowel variation quality (the first, second and third formants, hereafter F1, F2 and F3) within L1 and L2 Arabic language. The objective is to put forward formant variation in vowel production in Modern Standard Arabic (MSA) spoken by native vs. non-natives speakers. Thus, we examined the foreign accent of speakers within connected sentences produced by Arabic and American participants.

Afterward, we compared our results with those found in some Arabic dialect studies.

The paper is organized as follows. Section 2 exposes speech material and participants used in the study. Section 3 describes formants extraction. Section 4 shows experiments and findings. Section 5 gives the concluding remarks based on the analysis.

2 Materials

In this section, we outline materials used in the experimental setting. Information was given in detail about texts read, speakers, recordings and technical conditions. There are six vowels (three short vowels vs. three long vowels) and two diphthongs in Modern Standard Arabic (MSA) language. In our study, we focused only on formant frequencies as a part of acoustic features computed from short vowels (/a/, /u/ and /i/). Because vowels' formant frequencies can be impacted by phoneme co-articulation, recordings used were speech continuous sentences.

Totally 29 speakers participated in the experiment (15 natives/14 non-natives). They uttered five sentences taken from script 1 of the West Point corpus [18]. The English language is the mother tongue of non-native speakers. The first language of non-native speakers was the English language. Technical conditions of recording were as follows: normal speech rate, a sampling frequency of 22.05 kHz. A total of 145 recordings were used in the analysis. Table 1 shows the number and gender of speakers in the sample.

Table 1. Distribution of native and non-native speakers per gender

Native speakers		Non-native speakers	
Male	Female	Male	Female
5	10	6	8
Total	15	14	

Formants are distinctive frequency features that refer to frequency resonances. To compute speech formants, we annotated and segmented manually all speech material i.e. 145 recordings of the dataset onto their different segmental units (vowels and consonants) using Praat software [19]. Then, we extracted all vowels of the speech material, which constituted our speech signal material.

3 Feature Extraction

All speech signals of the vowel's dataset were first pre-emphasized with a pre-emphasis filter that is used to enhance the high-frequency components of the spectrum. This is performed by applying the following formula:

$$x_n' = x_n - ax_{n-1} \tag{1}$$

where $a\psi$ is the pre-emphasis coefficient which should be in the range $0 \le a < 1$.

Hamming window function was applied on each frame of the signal, to reduce boundary effects. The impulse response of the Hamming window is defined as follows:

$$(n) = 0.54 - 0.46cos(2\pi n/(N-1)) \tag{2}$$

where $0 \le n \le N - 1$.

The normal frame length was about 30 ms to ensure stationary property

Most formant tracking algorithms are based on Linear Predictive Coding (LPC). LPC coefficients are extracted from pre-processed speech. The estimated future speech samples from a linearly weighted summation of past p-samples using the method of least squares is

$$\hat{x}(n) = -\sum_{k=1}^{p} a(k)x(n-k) \tag{3}$$

where $x(n)$ and $\hat{x}(n)$ are speech samples and their estimates and

$a(k) = [a(1), a(2), \ldots a(p)]T$ is the LPC parameters and p is the linear predictive (LP) filter order.

4 Results and Discussions

Formants refer to resonant frequencies of the vocal tract that appear as clear peaks in the speech spectrum. Formants display a large concentration of energy within voiced phonemes as vowels. Most often, the two first formants, F1 and F2, are sufficient to identify the kind of vowel. Nevertheless, in the study, we performed three formants (F1, F2 and F3) values for each vowel to show variation in pronunciation between native and non-native speakers.

The first investigation seeks to provide the assessment of the mean values of all formants for both corpora categories i.e. short vowels produced by native and non-native MSA speakers groups. The calculation was conducted regardless gender of speakers (male/female). Table 2 outlines in detail the average and standard deviation values of F1, F2 and F3 measured for each short vowel /a/, /u/ and /i/.

Table 2. Means and standard deviation of F1, F2 and F3 (Hertz) of native (L1) and non-native (L2) speakers

		/a/	std	/u/	std	/i/	std
F1	L1	756,50	125,39	456,54	130,13	435,19	116,58
	L2	722,95	130,50	544,68	200,89	487,36	217,41
F2	L1	1643,09	244,87	1284,66	441,14	2139,58	357,23
	L2	1514,58	264,37	1420,32	472,96	2089,83	368,68
F3	L1	2735,39	325,26	2712,75	253,23	2929,33	289,58
	L2	2702,39	275,37	2785,36	356,50	2819,84	301,39

Figure 1 displays formants measures computed from L1 and L2 MSA for each short vowel /a/, /u/ and /i/. When comparing the L1 vowels against their L2 counterparts, we can observe that the averages of F1, F2, and F3 calculated in /a/ produced by the non-natives were less compared to their homologs in native speech. In contrast, for /u/ uttered by L2 participants, the values of F1, F2, and F3 increased compared to /u/ produced by Arabic speakers. In /i/, a slight variation was noticed in the second formant between both groups. F1 of native speakers was less compared to those of non-native speakers. In the case of F3, natives gave higher value.

Figure 2 illustrates the plot of the first two formants computed of both groups. The chart shows that the short vowels of non-native speakers are positioned slightly more central than their native counterparts: this tendency is more outspoken for /u/ and /i/. As it can be seen, the vowel diagram plotted for native speakers is larger than the triangle obtained by the second group. The diagram shows also that the latter is almost located inside the triangle drawn for native subjects. Statistical analysis indicates a significant effect in L2 pronunciation in both F1 and F2 formant for the vowel /a/. Likewise, results of /u/ and /i/ show also a significant effect on F1 for both vowels.

Comparative analysis was also carried out between the outcomes of this study and some other previous works [14–16]. Indeed, we compared our results with those obtained from researches conducted on Arabic dialects (7 and 1 dialects respectiely) by [14] and [15] and another study realized on Modern Standard Arabic (MSA) [16]. The Arabic dialect accents are the Qatari, the Lebanese, the Saudi, the Tunisian, the Syrian, the Sudanese, the Jordanian and the Egyptian. The authors built their experimental studies basing on a different kind of corpus (syllable, words, and connected speech).

Figure 3 depicts the location of MSA native and non-native vowels between all different dialect pronunciations. As it can be seen from Fig. 3, all vowels /u/ of Arabic dialects are positioned outside the MSA vowel diagram. However, almost all dialect values of /a/ and /i/ are within the MSA triangle. The chart shows also, that /u/ of non-native speakers is more central than all other dialects 'values. Regarding, /a/ and /i/ of the same group the plot displays that almost localized in the periphery of the graph.

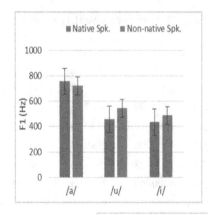
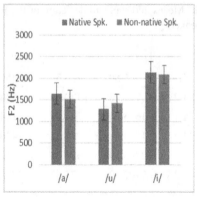

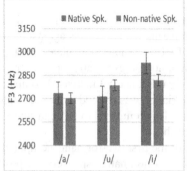

Fig. 1. Comparison of formant averages for both native and non-native MSA short vowels

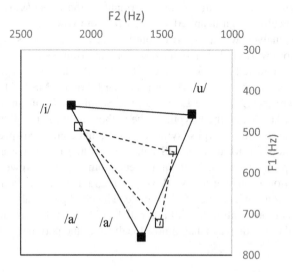

Fig. 2. Vowel diagram of native vs. non-native speakers

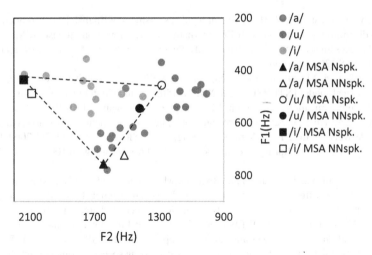

Fig. 3. Comparison of MSA native and non-native vowels with dialect vowels

5 Conclusion

The study deals with the formant analysis of Modern Standard Arabic (MSA) vowels produced by15 native (L1) and 14 non-natives (L2) speakers. Three formants were computed from all short vowels (/a/, /u/ and /i/) of the speech material. We computed for all vowels the three first formants F1, F2 and F3. When comparing the L1 vowels against their L2 counterparts, results show that formants calculated in /a/ produced by the non-natives were less compared to their homologs in native speech. In contrast, for /u/ uttered by L2 participants, the values of F1, F2, and F3 increased compared to /u/ produced by Arabic speakers. Regarding the vowel /i/, a slight variation was noticed in the second formant between both groups. The vowel diagram plotted from F1 and F2 values for both native and non-native speakers show that the short vowels of L2 speakers are positioned slightly more central than their native counterparts. This tendency is more outspoken for /u/ and /i/. Non-native triangle is almost located inside the triangle drawn for native subjects.

A comparative analysis of formants values was also carried out between native and non-native MSA with results of some eight Arabic dialect studies. The findings show that that /u/ of non-native speakers is more central than all other values computed in Arabic dialects. Concerning /a/ and /i/ uttered by L2 participants, the outcomes show that the latter are localized in the periphery of the vowel diagram.

References

1. MacDonald, M.: The influence of Spanish phonology on the English spoken by United States Hispanics. In Bjarkman, P., Hammond, R. (eds.) American Spanish pronunciation: Theoretical and applied perspectives, Washington, D.C.: Georgetown University Press, pp. 215–236, ISBN 9780878404933. (1989)

2. Nicolao, M., Beeston, A.V., Hain, T.: Automatic assessment of English learner pronunciation using discriminative classifiers. IEEE International Conference on Acoustics, Speech and Signal Processing (ICASSP), Brisbane, QLD, (2015). pp. 5351–5355, doi: https://doi.org/10.1109/ICASSP.2015.7178993.
3. Marzieh Razavi, M., Magimai Doss M.: On recognition of non-native speech using probabilistic lexical model INTERSPEECH 201. In: 15th Annual Conference of the International Speech Communication Association, 14–18 September, Singapore (2014)
4. Alotaibi, Y.A., Hussain, A.: Speech recognition system and formant based analysis of spoken Arabic Vowels. In: Lee, Y.-H., Kim, T.-H., Fang, W.-C., Ślęzak, D. (eds.) FGIT 2009. LNCS, vol. 5899, pp. 50–60. Springer, Heidelberg (2009). https://doi.org/10.1007/978-3-642-105 09-8_7
5. Droua-Hamdani, G., Selouani, S.A., Boudraa, M.: Speaker-independent ASR for modern standard Arabic: effect of regional accents. Int. J. Speech Technol. 15(4), 487–493 (2012)
6. Droua-Hamdani, G., Sellouani, S.A., Boudraa, M.: Effect of characteristics of speakers on MSA ASR performance. In: IEEE Proceedings of the First International Conference on Communications, Signal Processing, and their Applications (ICCSPA 2013), pp. 1–5. (2013).
7. Droua-Hamdani, G.: Classification of regional accent using speech rhythm metrics. In: Salah, A., Karpov, A., Potapova, R. (eds) Speech and Computer. SPECOM 2019. Lecture Notes in Computer Science, vol 11658, pp. 75–81. Springer, Cham (2019)
8. Droua-Hamdani, G.: Formant frequency analysis of msa vowels in six algerian regions. In: Karpov, A., Potapova, R. (eds.) SPECOM 2020. LNCS (LNAI), vol. 12335, pp. 128–135. Springer, Cham (2020). https://doi.org/10.1007/978-3-030-60276-5_13
9. Farchi, M., Tahiry, K., Soufyane, M., Badia, M., Mouhsen, A.: Energy distribution in formant bands for Arabic vowels. Int. J. Elect. Comput. Eng. 9(2), 1163–1167 (2019)
10. Mannepalli, K., Sastry, P.N., Suman, M.: analysis of emotion recognition system for Telugu using prosodic and formant features. In: Agrawal, S.S., Dev, A., Wason, R., Bansal, P. (eds.) Speech and Language Processing for Human-Machine Communications. AISC, vol. 664, pp. 137–144. Springer, Singapore (2018). https://doi.org/10.1007/978-981-10-6626-9_15
11. Korkmaz, Y., Boyacı, A.: Classification of Turkish Vowels Based on Formant Frequencies. In: International Conference on Artificial Intelligence and Data Processing (IDAP), pp. 1–4, Malatya (2018). https://doi.org/10.1109/IDAP.2018.8620877
12. Natour, Y.S., Marie, B.S., Saleem, M.A., Tadros, Y.K.: Formant frequency characteristics in normal Arabic-speaking Jordanians. J. Voice 25(2), e75–e84 (2011)
13. Abd Almisreb, A., Tahir, N., Abidin, A.F., Md Din, N.: Acoustical Comparison between /u/ and /u:/ Arabic Vowels for Non-Native Speakers. Indonesian J. Elect. Eng. Comput. Sci. 11(1), 1–8 (2018). https://doi.org/10.11591/ijeecs.v11.i1.
14. Rusza, J., Cmejla, R.: Quantitative acoustic measurements for characterization of speech and voice disorders in early-untreated Parkinson's disease. J. Acoust. Soc. Am. 129, 350 (2011). https://doi.org/10.1121/1.3514381
15. Abou Haidar, L.: Variabilité et invariance du système vocalique de l'arabe standard, Unpubl. PhD thesis, Université de Franche-Comté. (1991).
16. Alghamdi, M.: A spectrographic analysis of Arabic vowels: a cross-dialect study. J. King Saud Univ. 10(1), 3–24 (1998)
17. Newman, D.L., Verhoeven, J.: Frequency Analysis of Arabic Vowels in Connected Speech Antwerp papers in linguistics, Vol. 100, pp. 77–87 (2002)
18. Linguistic Data Consortium LDC. http://www.ldc.upenn.edu.
19. Boersma, P., Weenink, D.: Praat: doing phonetics by computer. http://www.praat.org. Accessed Mar 2010

Telecom and Artificial Intelligence

A New Strategy for Deploying a Wireless Sensor Network Based on a Square-Octagon Pattern to Optimizes the Covered Area

Absa Lecor[1]([⊠]), Diery Ngom[1], Mohamed Mejri[2], and Senghane Mbodji[1]

[1] Alioune Diop University of Bambey, Bambey, Senegal
{absa.lecor,diery.ngom,senghane.mbodji}@uadb.edu.sn
[2] Laval University, Quebec City, Canada
mohamed.mejri@ift.ulaval.ca

Abstract. In recent years, sensor networks have grown exponentially and they are used in many fields of applications such as the monitoring of environment parameters, intelligent agriculture, surveillance of area, smart city, monitoring biological parameters of patients, etc. A Wireless Sensor Network (WSN) is a set of sensors deployed over a geographical area so that each node can collect data from the environment, do local processing and transmit them to a sink node or base station using multi-path routing. An optimal deployment of sensors in the area of interest is required for the network to be efficient. So a good deployment can ensure a good coverage of the area, a better network connectivity and also an energy saving. In the literature several methods for deploying sensors are proposed. In this paper, we work on strategies based on the grid. We propose to divide the area into a square-octagonal pattern. The pattern consists of a square polygon and two octagon polygons. Based on our method, we evaluate the number of sensors used for the deployment of the sensors network, the percentage of coverage obtained, and then we make comparisons with other proposals in the literature. The obtained results showed that our proposal enable to ensure 91.03% of the coverage area compared to other strategies of sensors deployment.

Keywords: Square-octagon pattern · Coverage · Deployment strategy · Wireless sensors network

1 Introduction

Wireless sensor networks are an emerging technology that is becoming increasingly popular in both the civilian and military domains. A WSN allows sensors to detect information (such as temperature or pressure) about the environment in which they are deployed, and then send it to collection points, called "base stations". This is how WSN are used in several domains such as environmental monitoring, intelligent transportation, building, patient monitoring, agriculture, etc. In many applications, the deployment of sensors is done without prior

Y. Faye et al. (Eds.): CNRIA 2021, LNICST 400, pp. 61–76, 2021.
https://doi.org/10.1007/978-3-030-90556-9_6

planning. However, an optimal deployment of sensors in the area of interest is necessary for the network to be effective. Thus, a good deployment can provide good coverage of the area, better connectivity of the network and also energy savings. It is in this context that we will conduct our studies to ensure a very good coverage with an optimal number of sensors. In the literature, several sensor deployment methods have proposed. In this paper, we work on a grid-based strategy. It consists in dividing the area into a square-octagonal pattern. The pattern consists of one square polygon and two octagonal polygons. Based on our method, we evaluate the number of sensors used for the deployment, the percentage of coverage obtained, and then we compare it with other approaches in the literature presented in Table 4 of Sect. 3 and Fig. 9 of Sect. 4.

The rest of this paper is structured as follows. Section 2 gives some related works on the basic concepts of coverage and connectivity in a sensor network. Sections 3 we present in detail the new WSN deployment technique based on a square-octagon pattern to Optimizes the Covered Area. Section 4 evaluates the approach. We evaluate the coverage efficiency (%) obtained by our deployment pattern to provide partial area coverage with n-sided polygon models. We also compare for each pattern of deployment, the ratio between the sensing radius and the communication radius. Some concluding remarks and future work are given in Sect. 5.

2   Related Works

Recently, many researchers have studied sensor deployment techniques in WSN. Authors in [1–7] have worked on the basic concepts of coverage and connectivity in a sensor network. In the following, we show the different types of sensor deployments in sensors network.

2.1   Sensors Classification

A WSN consists of a set of nodes which can be deployed randomly or deterministically over a given interest area. Each sensor node has a coverage radius, a communication radius, an energy that represents its lifetime. Sensors can also be differentiated by their mobility. We have two types of sensors according to this criterion: fixed sensors and mobile sensors. In [2], the author compares these two types of sensors in terms of energy consumption and they conclude that mobile sensors consume more energy than fixed ones. Indeed, the movement of a sensor can have a direct effect on the efficiency of the coverage area. On the other hand, the author in [8] shows that mobile nodes allow closing holes in the network.

2.2   Types of Coverage in WSN

Coverage is an important performance metric in WSN, which reflects how well a sensing field is monitored. Its purpose is to collect good information of the area of interest [8]. Depending on the position of the sensors, we can have several types of coverage [1,9] such as barrier coverage, point coverage and area coverage.

- **Barrier coverage:** In this case, the objective is to achieve an arrangement of sensors with the task of maximizing the detection probability of a specific target penetration through the barrier.
- **Point coverage:** The goal of this technique is to cover a set of points (target) with known positions that need to be monitored. This coverage scheme focuses on determining the exact positions of sensor nodes while guaranteeing and efficient coverage for a limited number of immobile targets. Depending on the mobility of the points, we have two types of coverage: fixed points and mobile points.
- **Area coverage:** The main objective is to cover (monitor) a region and to maximize the detection rate of a specific area.

In the following, we define analytically the point coverage and the area coverage techniques. Let $M = \{S1, S2, ..., Sn\}$ be a set of sensors nodes and A denoted of a given area. On the other hand, let r_s represent the sensing range of a sensor node in the WSN. In this case, a sensor $Si \in M$ cover a point $q \in A$ if and only if: $d(Si, q) \leq r_s$. While $d(Si, q)$ represent the Euclidean distance between the sensor node Si and the point q. The coverage of area by a sensor $Si \in M$, denoted by $C(Si)$, is defined by:

$$C(Si) = \{d \in A \mid d(Si, q) \leq r_s\}$$

Finally, the area coverage by a set of sensors $Mn = \{S1, S2, ..., Sn\}$, denoted by $C(Mn)$ is defined by:

$$C(Mn) = \sum_{k=1}^{|Mn|} C(Sk)$$

In this paper, we focus on the area coverage technique. Most of the previous studies on this problem have focused on how to reduce the number of sensors to cover an area. In our case, we try to maximize the coverage of an area with a given number of sensors having the same characteristics.

2.3 Types of WSN Deployment

In order to ensure a good coverage of an area together with a network connectivity in the WSN, it is important to study and take into consideration the existing sensors deployment strategies. There are two types of sensor deployment techniques: random and deterministic [8,9]. For the first case, it consists in knowing the exact position of the sensor before its deployment, contrary to the second deployment technique. Otherwise, sensor deployment strategies depend on the applications. Some sensor network applications require a full-area coverage of the region and others require a partial or a target coverage. For example, critical area monitoring and smart agriculture requires full coverage of the area.

2.4 Sensors Deployment Algorithm in WSN

Deployment can be uniform or non-uniform in a network architecture that can be centralized or distributed.

Deployment by the Meta-heuristic Method: In [7], the authors proposed a random sensor deployment with a meta-heuristic approach. The sensors are dynamic over time to ensure robustness of the network in terms of coverage. To achieve this objective, each sensor is programmed to participate dynamically on the performance of the network. The data fusion technique is used through clusters. Sensor deployment can also be done using a small aircraft. In [7], the deployment takes three phases: the predeployment, the postdeployment and the redeployment. The authors have applied their approach to different cases: area coverage, barrier coverage and point coverage. For a better efficiency of the heuristic method in relation to the virtual forces, a genetic algorithm is presented in [9]. Here, the authors show that the heuristic method is more efficient in terms of coverage and the number of used sensors.

Virtual Force Algorithms: This algorithm uses a simple deployment. Between two sensors, a force is exerted on one according to its distance from its neighbor and a fixed threshold. In [1], the authors explain in detail the execution of the different virtual forces as follows.

- If the distance between two sensors is less than the threshold, they exert a repulsive force on each other.
- If the distance between two sensors is greater than the threshold, they exert an attractive force on each other.
- If the distance between the two sensors is equal to set threshold, they exert a zero force on each other.

Figure 1 gives the three types of forces where the radius of the circle represents the threshold value.

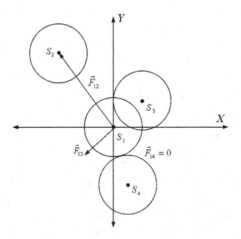

Fig. 1. Virtual forces [3]

In [3], the authors proposed an algorithm to follow a moving object with virtual forces. They have developed a probabilistic approach based on the information collected by a group of sensors deployed to monitor a moving object.

Strip Deployment Algorithm: In a given area, the sensors can be deployed either in a single horizontal strip or in two strips. It consists of placing a set of sensors in a plane where the distances between them are denoted by d_α and d_β as shown by Fig. 2. The columns are deployed either from left to right or from the other side. In [8], the authors make a physical deployment of the sensors in horizontal stripes and then they treat the notion of cooperative communication between them. In the study presented in [8] and [9], each sensor in the horizontal band carries a crucial information in order to ensure the coverage of the network. Then the authors make a comparison between the physical and information coverage of an area.

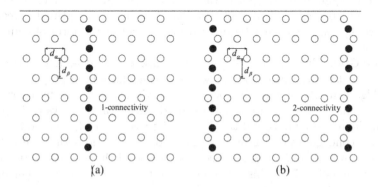

Fig. 2. Strip deployment in [8].

Computational Geometry Based Strategy: The geometric strategies are based on objects such as points, segments, lines, polygons, etc. According to [1] and [8], the two most used methods in WSN are the Voronoi diagram and the Delaunay triangle which are based on irregular patterns as shown by Fig. 3.

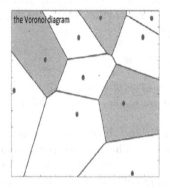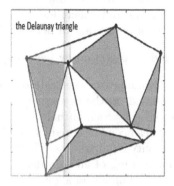

Fig. 3. Computational geometry based strategy in [1]

Grid-Based Strategy: The grid-based strategy ensures a deterministic deployment. In this type of deployment, the sensor position is fixed according to a chosen grid pattern. In [10], the grids or blocks can be all identical regular polygons placed side by side or grids which can be regular and nonidentical polygons placed side by side. Many works have focused on the study of grids for an optimal deployment of sensors. Among them, we have:

- Triangular Grid: The zigzag-shaped sensor deployment is presented in [11] with a distance of $\sqrt{3}r_s$ (where r_s is sensing range of a sensor) between two adjacent nodes over an angle of 60°. The author provides an algorithm that ensures 91% complete coverage and 9% redundancies. In [1] it has been shown that the equilateral triangle is the polygon with side equal to 3 provides maximum coverage in terms of area. Then, to have an optimal percentage of coverage, an area must be divided into a grid in the form of equilateral triangles.
- Square Grid: The author of [12] cuts his work area into square grids for the deployment of his sensor nodes. Two algorithms have been proposed:
  - Algorithm Version 1: It consists in dividing the area into rows and columns. A first group of sensors are deployed on the rows and columns without overlapping and the distance between two adjacent nodes is $2r_s$. A second group of sensors are placed on intersections that are not covered by the first group of sensors, at this level the distance between two neighbors is $0.7r_s$. This version of algorithm gives a coverage percentage of 78% of the area.
  - Algorithm Version 2: It consists in dividing the area into rows and columns. The sensors are deployed on the rows and columns with overlapping. The distance between two nodes is $0.7r_s$ along the vertical axis and also along the horizontal axis. This version of algorithm gives a coverage percentage of 73% of the area.
- Hexagonal grid: This model consists in dividing the area into cells of hexagonal shape. This model has been presented in [13] with a distance between sensors of $d \leq \sqrt{3}r_s$.
  In [14], the authors compare different uniform and identical grid-shaped deployments and concludes that the hexagon provides better performance in terms of percentage of network coverage.

We conclude this section by two comparative tables:

- Table 1 summarizes the state of the art. It presents all the approaches that have worked on the deployment of sensor networks with the different techniques used by each author.
- Table 2 has been presented by Mahfoudh in [1] and shows the coverage and the connectivity requirements of WSN applications. Thus, to ensure the collection and monitoring of climate data, we choose to deploy on a partial coverage $\leq 80\%$ based on a grid algorithm. The cutting pattern of our algorithm is the square-octagon (two octagons and one square).

Table 1. Summary of the State of the Art

Approach	Application	Deployment	Coverage	Quality	Weakness
[3]	Monitoring	Random	Virtual force	Auto-replace	Energy
[13–16]	Monitoring	Random	Heuristic	Evolutionary	Energy
[1, 17]	Monitoring	Deterministic	Geometric	Stable	Not uniform
[8, 18]	Monitoring	Deterministic	Strip	Precision	High cost
[19]	Monitoring	Determ./random	Grid pattern	Auto-replace	Energy
[10, 11, 20]	Monitoring	Determ./random	Grid pattern	Accuracy	Choice pattern

3 Approach

In this section, we will present in detail our approach.

Table 2. Applications of WSN

Applications	Type coverage	Coverage	Connectivity	Response time	Mobility
Industrial site	Area coverage	Full	Permanent	Temporal	Fixed
Fire detection	Area coverage	Partial	Permanent	Threshold	Fixed
Intruder detection	Barrier coverage	Full	Permanent	Presence	Fixed
Border surveillance	Barrier coverage	Full	Permanent	Presence	Fixed
Air pollution	Area coverage	Partial	Intermittent	Temporal	Mob./Fix.
Patient monitoring	Point of interest	Full	Permanent	Real time	Fixed
Climate monitoring	Area coverage	Partial	Intermittent	Temporal	Mob./Fix

3.1 Preliminaries

Some useful preliminary parameters and notations related to a polygon (example Fig. 4) are presented below:

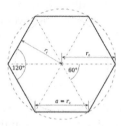

Fig. 4. Polygon of n sides (n = 6)

- We denote by $Q(n, a)$ a polygon of n sides of length a each.
- The perimeter of $Q(n, a)$ is $P(Q(n, a))$ and its area is $A(Q(n, a))$. We simply write P and A when $Q(n, a)$ is clear from the context.
- We denote by r_i the measure of the apothem (the radius of the biggest inscribed circle) and at the same time r_i is the sensing radius of the sensor (r_s).
- We use r_c to denote the radius of the circumscribed circle and at the same time r_c is the communication radius of the sensor.

The following equations give the relationship between the different parameters of a polygon.

$$a = 2 \times r_c * \sin(\frac{\pi}{n}) = 2 \times r_i * \tan(\frac{\pi}{n})$$

$$r_i = r_c \times \cos(\frac{\pi}{n}) = (1/2) \times a \times \cot(\frac{\pi}{n})$$

$$r_c = \frac{a}{2 \times \sin(\frac{\pi}{n})} = \frac{r_i}{\cos(\frac{\pi}{n})}$$

$$A = \frac{n \times a^2}{4 \times \tan(\frac{\pi}{n})}$$

$$P = a \times n$$

In the harmony of the world of KEPLER in 1619, the way in which an area is divided is called paving. It is shown in [10] that there are two ways of dividing an area to reach a paving. Either one can divide the zone into identical convex regular polygons with a common vertex P and whose sum of their angles at this point P is 360° which is the regular paving, or into a convex regular polygon and not identical (called a semi-regular paving).

We propose to use the grid-based algorithm with the principle of semi-regular paving. Firstly, we calculate the area of the zone and then depending on the communication radius and the catchment radius, we set the size of each dimension of our grid pattern. Secondly, the area of interest will be divided into two octagons and one square, then we place on the center of each octagon of our pattern an r_c communication beam sensor. Finally, we are going to look for the number of sensors that will be used and the percentage coverage of the area.

3.2 Hypothesis

The following assumptions are used by the approach proposed in this paper:

- Each sensor is omnidirectional, and covers an angle of 360°;
- Each node is considered as a disk with communication radius r_c and sensing radius r_s;
- All cutting work is based on the 2-dimensional space;
- Each pattern may have one or more polygons consisting of n dimensions of size a;

- The r_c of the sensor is equivalent to the radius of the circle circumscribed to the polygon;
- The r_s of the sensor is equivalent to the radius of the circle inscribed at the polygon.

3.3   Objectives

The main goal of this approach is to ensure an optimized coverage of the city of Bambey and the Ferlo area in Senegal based on a new technology of sensor network deployment. This technique is based on a cutting of the area in a square-octagon pattern in order to optimize the number of sensors for a better coverage and to ensure connectivity and robustness of the network. Thus to achieve this objective we need to achieve a set of well defined sub-objectives:

1. Optimal number of sensors;
2. Better coverage percentage;
3. Better sensor locations and
4. Network connectivity and robustness.

The following table clearly shows our different working criteria set to reach our goal. On a set of approaches, we are going to compare the coverage efficiency, the cost of the installation, the energy used by a node, the distance between two nodes and also the mobility of the sensors between different patterns. The number of sensors is expressed as a function of r_c. It is equal to A_z (the area of the total zone) divided by the A_c (area of a cell) expressed as a function of r_c. The distance d between two nodes is equal to twice the coverage radius which is also equal to twice the radius of the circle inscribed to the cell.

$$\texttt{number\_of\_nodes} = \frac{A_z}{A_c}$$

Table 3 compares the coverage efficiency, the cost of the installation, the energy used by a node, the distance between two nodes and also the mobility of the sensors between different patterns.

The value of d_1 gives the distance between two nodes along the horizontal axis and d_2 is the distance between two nodes along the vertical axis.

3.4   Algorithm

The proposed approach consists in dividing an area into a pattern containing two octagons and a square each (sometime denoted by $8^2 4$). The size of each side is a as shown by Fig. 5.

Table 3. Coverage efficiency comparison

Approach	Grid pattern	Coverage	Number of nodes	Distance between two nodes (d)
[11]	Triangle	39–41%	$\frac{A_z}{1.299r_c^2}$	$0.7r_s$
[12]	Square	73–78%	$\frac{A_z}{2r_c^2}$	$d_1 = 2r_s, d_2 = 0.7r_s$
[18, 20, 21]	Hexagon	82.69%	$\frac{A_z}{2.59r_c^2}$	$\sqrt{3}r_s$
[8, 18]	Strip	100%	$\frac{r_c}{r_s} \geq 2.5$	$\min(r_c, \sqrt{3}r_s)$
[19]	Semi-random	80–95%	Fixed	Variable
Proposed approach	Square + Octagon	91.03%	$\frac{A_z}{3.82r_c^2}$	$1.84r_s$

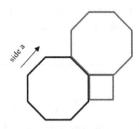

Fig. 5. Pattern of 2 octagons and 1 square

Patterns are connected to each others as shown in Fig. 6.

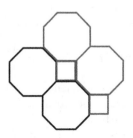

Fig. 6. Pattern connection

Once the complete zone is covered by patterns, it looks as shown by Fig. 7. After that sensors are positioned in the center of octagons and labeled as following:

- The BS station is center of the zone, denoted by label 0.
- Sensors having label $i + 1$, are those having a distance $2 * r_c$ from a sensor having a label i.

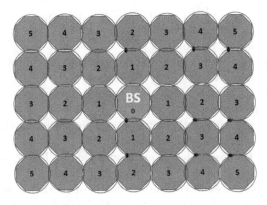

Fig. 7. Zone covering and sensor placement

For the sake of simplicity, we assume that area is a rectangle with a length l and width w. The center of the area will be as the position $(0,0)$ used as the origin of our repair. Algorithm 1 gives the positions of sensors and their labels. Positions will be returned in table called *Positions* where $Positions[i,j]$ gives x_i (abscissa of the i-th sensor to the right if $i > 0$, to the left if $i < 0$), y_j (ordinate of j-th sensor up if $j > 0$ and down if $j < 0$) and $h_{i,j}$ (the label that reflects the number of hops to reach the BS). At position $(0,0)$ we have the BS.

Algorithm 1: Sensor Positioning and Labeling

Data: (l, w), r_c and n

Result: *Positions*

1 initialization;

2 $a = 2 * r_c * sin(\frac{\pi}{8})$;

3 $A_z = l * w$;

4 $A_{octagon} = \frac{n*a^2}{4*tan(\frac{\pi}{n})}$;

5 $A_{square} = a * a$;

6 $A_{pattern} = 2 * A_{octagon} + A_{square}$;

7 $Nbr_{patterns} = \lceil \frac{A_z}{A_{pattern}} \rceil$;

8 $Nbr_{sensors} = 2 * Nbr_{patterns}$;

9 $Position[0][0] = (x_{bs} = \frac{l}{2}, y_{bs} = \frac{w}{2}, 0)$;

10 **for** $(i = 0; i < \frac{l}{2*r_c}; i++)$ **do**

11   **for** $(j = 0; j < \frac{w}{2*r_c}; j++)$ **do**

12          $Positions[i][j] = (x_{bs} + i * 2 * r_c, y_{bs} + j * 2 * r_c, i + j)$;

13          $Positions[i][-j] = (x_{bs} + i * 2 * r_c, y_{bs} - j * 2 * r_c, i + j)$;

14          $Positions[-i][j] = (x_{bs} - i * 2 * r_c, y_{bs} + j * 2 * r_c, i + j)$;

15          $Positions[-i][-j] = (x_{bs} - i * 2 * r_c, y_{bs} - j * 2 * r_c, i + j)$;

16   **end**

17 **end**

The following lines comment line by line the algorithm

line 1: initialization of variables;

line 2: calculate the value a as a function of the communication radius r_c of the size of the side of a polygon of size 8 sides;

line 3: calculate the area of the working area in rectangle is equal to length l multiplied by the argeur w;

line 4: calculate the area of the octagon as a function of a the size of a side. The octagon is a polygon of 8 sides $n = 8$;

line 5: calculate the area of the square with side a. The square is a polygon with four sides $n = 4$;

line 6: calculate the area of the pattern made of two octagons of a square;

line 7: calculates the number of patterns that can be cut out of the area of interest by dividing the area of the area by the area of a pattern;

line 8: calculates the number of sensors in the area of interest by multiplying the number of patterns by two. For each pattern we can place two sensors the center of each octagon;

line 9: gives us the position of the base station in our frame. The base station is in the center of the frame at position (0,0) and its position in the area of interest is (length /2, width/2) $(x_{bs} = \frac{l}{2}, y_{bs} = \frac{w}{2}, 0)$;

line 10: the index i represents the x-axis of a sensor. i traverses the area of interest from 0 to $(length/2r_c)$ $(\frac{l}{2*r_c})$;

line 11: the index j represents the y-axis of a sensor. j traverses the area of interest from 1 to $(width/2r_c)$ $(\frac{w}{2*r_c})$;

line 12: represents the position (i, j) of a sensor;

line 13: represents the position $(i, -j)$ of a sensor;

line 14: represents the position $(-i, j)$ of a sensor;

line 15: represents the position $(-i, -j)$ of a sensor;

line 16: the end of the loop for with index i;

line 17: the end of the loop for with index j.

here is an example of application of the algorithm:
The surface of the working area is 500 m² with length 25 m and width 20 m and the communication radius of the sensors is 4 m.

1   $a = 2 * r_c * sin(\frac{\pi}{8}) \Rightarrow a = 2 * 4 * sin(\frac{\pi}{8}) = 3.1m;$

2   $A_z = 500m^2;$

3   $A_{octagon} = \frac{n*a^2}{4*tan(\frac{\pi}{n})} \Rightarrow A_{octagon} = \frac{8*3^2}{4*tan(\frac{\pi}{8})} = 45.25m^2;$

4   $A_{square} = a * a \Rightarrow A_{square} = 4 * 3.1 = 9.37m^2;$

5   $A_{pattern} = 2 * A_{octagon} + A_{square} \Rightarrow A_{pattern} = 2 * (45.25) + 9.37 = 99.88m^2;$

6   $Nbr_{patterns} = \lceil \frac{A_z}{A_{pattern}} \rceil \Rightarrow Nbr_{patterns} = \lceil \frac{500}{99.88} \rceil = 5patterns;$

7   $Nbr_{sensors} = 2 * Nbr_{patterns} \Rightarrow Nbr_{sensors} = 2 * 5 = 10sensors;$

8   $Position[0][0] = (x_{bs} = \frac{l}{2}, y_{bs} = \frac{w}{2}, 0) \Rightarrow Position[0][0] = (x_{bs} = \frac{25}{2}, y_{bs} = \frac{20}{2}, 0);$

9   **for** $(i = 0; i < \frac{25}{2*4}; i++)$ **do**

10   **for** $(j = 1; j < \frac{20}{2*4}; j++)$ **do**

11        $Positions[i][j] = (x_{bs} + i * 2 * 4, y_{bs} + j * 2 * 4, i + j);$

12        $Positions[i][-j] = (x_{bs} + i * 2 * 4, y_{bs} - j * 2 * 4, i + j);$

13        $Positions[-i][j] = (x_{bs} - i * 2 * 4, y_{bs} + j * 2 * 4, i + j);$

14        $Positions[-i][-j] = (x_{bs} - i * 2 * 4, y_{bs} - j * 2 * 4, i + j);$

Table 4 gives the results obtained by applying Algorithm 1 on an area of size 500 m^2 and with a radius of 4 m. The result is compared to other polygon pattern of dimension n.

4   Evaluation and Performance

In this section, we evaluate the coverage efficiency (%) obtained by our deployment pattern to provide partial area coverage with n-sided polygon models. We also compare for each pattern of deployment, the ratio between the sensing radius and the communication radius. The ratio between the coverage radius and the communication radius allows us to know the predefined redundancy capacity of the network. The lower this ratio is, the lower the redundant zones are and the more the ratio tends towards zero, the more the redundancy increase as it is shown in Table 5.

Table 4. Coverage and efficiency

n	Cell area	Number of sensors	Theoretical coverage	Efficiency
3	20.78	24	1209.25	41.34
4	32	16	785.06	63.68
5	38	13	660.91	75.65
6	41.56	12	604.62	82.69
7	43.68	11	574.97	86.96
8	45.25	11	554.87	90.11
9	46.28	11	545.32	91.68
10	46.96	11	534.72	93.49
$8^2 4$	99.879	10	549.25	91.03

Table 5. Grid algorithm, efficiency and coverage

Grid pattern	Efficiency	Coverage	r_s/r_c
Triangle	39–41%	Full	≤0.65
Square	73–78%	Full	0.7
Hexagon	82.69%	Full	0.866
Strip	100%	Full	Variable
Semi-random square	80–95%	Partial	Variable
1square-2octogon	91.03%	Partial	0.9238

Figures 8 and 9 show respectively the number of sensors required for an area of size 500 m², and the percentage of coverage obtained by our deployment pattern compared to others (triangle, square, hexagon).

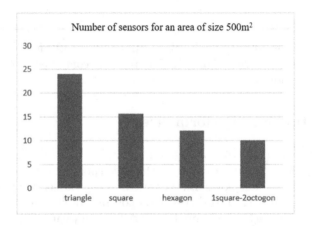

Fig. 8. Number of sensors required depending on the pattern of deployment

As shown in Fig. 8, the number of sensors required to cover an area of size 500 m² is much lower with the 1Square-2octagon pattern compared to the hexagon, square and triangle. As illustrated in Fig. 8, our deployment pattern allows to decrease the number of sensors, which optimizes the cost.

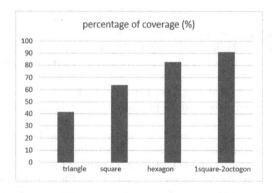

Fig. 9. Percentage of coverage

In Fig. 9, we compare areas of identical size that are partitioned into triangles, squares, hexagons and our pattern (1square-2octagon). We find that our pattern has a better percentage of coverage compared to other patterns. The working area is 500 m^2 and the radius used to calculate the results is 4 m.

5   Conclusion

This paper proposes a new algorithm for sensor deployment that minimizes the number of sensors and significantly improves surface coverage in WSN. Additionally, we investigated other grid-based deployment models such as triangles, squares, and hexagons. The comparisons of our pattern with those of the literature allowed to conclude that our algorithm gives better performances in terms of coverage area and number of sensors required to cover a given region.

As future work, we intend to use our pattern deployment in the areas of Bambey and Ferlo in Senegal for the implementation of an application for the collection, the processing and the transmission of climatic data in real time using sensor networks.

References

1. Mahfoudh, S., Minet, P., Laouiti, A., Khoufi, I.: Survey of deployment algorithms in wireless sensor networks: coverage and connectivity issues and challenges. IJAACS **10**(4), 341 (2017)
2. Bomgni, A.B., Mdemaya, G.B.J.: A2CDC: area coverage, connectivity and data collection in wireless sensor networks. NPA **10**(4), 20 (2019)
3. Zou, Y., hakrabarty, K.: Sensor deployment and target localization based on virtual forces. In: IEEE INFOCOM 2003. Twenty-second Annual Joint Conference of the IEEE Computer and Communications Societies (IEEE Cat. No.03CH37428), San Francisco, CA, vol. 2, pp. 1293–1303. IEEE (2003)
4. Liu, Y.: A virtual square grid-based coverage algorithm of redundant node for wireless sensor network. J. Netw. Comput. Appl. 7 (2013)

5. Liao, C., Tesfa, T., Duan, Z., Ruby Leung, L.: Watershed delineation on a hexagonal mesh grid. Environ. Model. Softw. **128**, 104702 (2020)
6. Firoozbahramy, M., Rahmani, A.M.: Suitable node deployment based on geometric patterns considering fault tolerance in wireless sensor networks. IJCA **60**(7), 49–56 (2012)
7. Mnasri, S., Nasri, N., Val, T.: The deployment in the wireless sensor networks: methodologies, recent works and applications. In: Conference Proceedings, p. 9 (2014)
8. Cheng, W., Lu, X., Li, Y., Wang, H., Zhong, L.: Strip-based deployment with cooperative communication to achieve connectivity and information coverage in wireless sensor networks. Int. J. Distrib. Sens. Netw. **15**(10) (2019)
9. Farsi, M., Elhosseini, M.A., Badawy, M., Ali, H.A., Eldin, H.Z.: Deployment techniques in wireless sensor networks, coverage and connectivity. IEEE Access **7**, 28940–28954 (2019)
10. Salon, O.: Quelles tuiles?! (Pavages apériodiques du plan et automates bidimensionnels). Journal de Théorie des Nombres de Bordeaux **1**(1), 1–26 (1989)
11. Hawbani, A., Wang, X.: Zigzag coverage scheme algorithm & analysis for wireless sensor networks. Netw. Protoc. Algorithms **5**, 19–38 (2013)
12. Hawbani, A., Wang, X., Husaini, N., Karmoshi, S.: Grid coverage algorithm & analysis for wireless sensor networks. Netw. Protoc. Algorithms **6**, 1–19 (2014)
13. Alexander, T.: La créativité et le génie ne peuvent s'épanouir que dans un milieu qui respecte l'individu et célèbre la diversité. p. 8
14. Senouci, M.R., Mellouk, A.: A robust uncertainty-aware cluster-based deployment approach for WSNs: coverage, connectivity, and lifespan. J. Netw. Comput. Appl. **146**, 102414 (2019)
15. Loscri, V., Natalizio, E., Guerriero, F., Mitton, N.: Efficient coverage for grid-based mobile wireless sensor networks. In: Proceedings of the 11th ACM symposium on Performance evaluation of wireless ad hoc, sensor, & ubiquitous networks - PE-WASUN 2014, Montreal, QC, Canada, pp. 53–60 ACM Press (2014)
16. Sharma, V., Patel, R.B., Bhadauria, H.S., Prasad, D.: Deployment schemes in wireless sensor network to achieve blanket coverage in large-scale open area: a review. Egypt. Inf. J. **17**(1), 45–56 (2016)
17. Iliodromitis, A., Pantazis, G., Vescoukis, V.: 2D wireless sensor network deployment based on Centroidal Voronoi Tessellation. p. 020009, Rome, Italy (2017)
18. Cheng, W., Li, Y., Jiang, Y., Yin, X.: Regular deployment of wireless sensors to achieve connectivity and information coverage. Sensors **16**(8), 1270 (2016)
19. Bomgni, A.B., Brel, G., Mdemaya, J.: An energy-efficient protocol based on semi-random deployment algorithm in wireless sensors networks. Int. J. Netw. Secur. **22**, 602–609 (2020)
20. Boukerche, A., Sun, P.: Connectivity and coverage based protocols for wireless sensor networks. Ad Hoc Netw. **80**, 54–69 (2018)
21. Li, R., Liu, X., Xie, W., Huang, N.: Deployment-based lifetime optimization model for homogeneous wireless sensor network under retransmission. Sensors **14**, 23697–23723 (2014)

Frequency Reconfigurable UHF RFID Dipole Tag Antenna Using a PIN Diode

Pape Waly Sarr[1,2(✉)], Ibra Dioum[1,2], Idy Diop[1,2,3], Ibrahima Gueye[1,2,3], Lamine Sane[1,2], and Kadidiatou Diallo[1,2]

[1] Faculté Des Sciences et Techniques, Ecole Supérieure Polytechnique (ESP), Université Cheikh Anta Diop, Dakar, Senegal
sarrpapewaly@gmail.com, {ibra.dioum,idy.diop, lamine.sane}@esp.sn, gueyeibrahimaig@gmail.com, diallokadidiatou06@gmail.com

[2] Laboratoire d'Informatique Télécommunications et Applications (LITA), Dakar, Senegal

[3] Laboratoire d'Imagerie Médicale et Bio-informatique (LIMBI), Dakar, Senegal

Abstract. This paper presents a design approach for a RFID reconfigurable dipole antenna for the UHF band. The frequency reconfiguration of this antenna, with a PIN diode, makes it possible to cover two frequency bands namely the European band (865.5–869 MHz) and the American band (902–928 MHz). The designed antenna is a copper dipole (e = 0.035 mm) printed on a FR4 epoxy substrate ε_r = 4.4 with a volume of $90 \times 25 \times 1.6$ mm^3. The reconfiguration was done by connecting a PIN diode to the ends of the two lines drawn at the top. The antenna has a 390 MHz (810–1200 MHz) bandwidth and covers the entire UHF RFID band. With reference to a 4w EIRP, the antenna can be read from 17 m. The simulations were done with HFSS which allowed us to find the antenna parameters, depending on the ON and OFF states of the PIN diode.

Keywords: RFID · PIN diode · UHF band · Tag antenna · Frequency reconfiguration

1 Introduction

With the development of RFID technology, UHF RFID (Ultra High Frequency 860–960 MHz) [1] is growing rapidly, notably thanks to the development of low cost liabilities Tags. The passive UHF tag technology with an emitted power of the order of 2 W makes it possible to reach a reading distance of about ten meters [2]. RFID systems are closely related to smart cards for storing data. These data are stored in an electronic device: the tag.

RFID antennas often encounter certain constraints such as capacity limitation, manufacturing cost, functionality in certain frequency bands. As a result, studies have been directed towards multiband functionality, significant miniaturization, the use of new dielectric materials and the strengthening of multifunctional capacities. In order to meet these needs, innovative methodologies are necessary for antennas design.

Y. Faye et al. (Eds.): CNRIA 2021, LNICST 400, pp. 77–86, 2021.
https://doi.org/10.1007/978-3-030-90556-9_7

Data transfer takes place via electromagnetic waves. Their applications make it possible to choose the frequencies of use, but our study will be limited to the UHF band (860–960 MHz). It is made up of two physical entities which are the tag and the RFID reader. In the UHF band, communication takes place by scattering electromagnetic waves through antennas on either side as shown in Fig. 1.

The RFID reader or interrogator consists of an RF module or base station (transmitter and receiver) which emits electromagnetic waves and receives the information sent by the tag through antennas in order to transmit them to the control device or host system. While the tag or receiver is composed of a chip which is adapted to an antenna which has the role of radiating an electromagnetic wave during the transmission of the data stored in the chip. This technique involves in contactless reading and writing of data in the non-volatile memory of the RFID tag through backscattered RF signals. The voltage that develops across the antenna terminal powers the chip and returns information to the reader by varying the input impedance and modulating the backscattered signals [8]. A very important factor in the design is the impedance matching which allows the tag to recover the maximum energy received by its antenna and transmit it to the chip.

In the literature, many RFID tag antennas are designed using different substrates and metallic track materials [7–13]. Different tag antennas are designed based on different chips available on the market to operate in the UHF frequency band [9–11]. Most of them can only reason on one frequency band. Some non-reconfigurable dual-band antennas were designed to cover the European and American RFID UHF bands as in [12]. However, the goal was to design a UHF RFID tag antenna that could operate in two frequency bands such as the European and American frequency bands by frequency reconfiguration using a PIN diode.

In this paper, we present a frequency reconfigurable miniature UHF RFID dipole antenna by insertion of a PIN diode. This antennas can operate in two UHF bands: European band (865–869 MHz) and American band (902–928 MHz). This antenna is made of a copper printed on an epoxy FR4 substrate (h = 1.6 mm and $\varepsilon_r = 4.4$) and is suitable for the Monza R6 chip. Simulations are done with the Ansys's electromagnetic simulation software named HFSS.

The remainder of the article is divided into a three sections: the second presents the design process of the sample (non reconfigurable) antenna and also the simulation results of the antenna. The Sect. 3 presents the simulation results of the reconfiguble antenna. The Sect. 4 concludes article.

2 Antenna Design Parameters

The design of the UHF tag antenna should have properties such as omnidirectional radiation, wide bandwidth and impedance matching with the chip to ensure maximum power transfer as shown in 1 [5].

$$Z_{ant} = Z_{chip}^* \tag{1}$$

Z_{ant} indicates the impedance of the UHF RFID tag antenna while, Z_{chip}^* is a complex conjugate of the input impedance of the chip [6]. Since RFID chips are often capacitive, the tag antenna must have an inductive impedance to achieve complex impedance matching. If the antenna parameters (gain, power transfer efficiency and wavelength) are Known, it is possible to calculate the reading range of the RFID tag antenna from 2 [6].

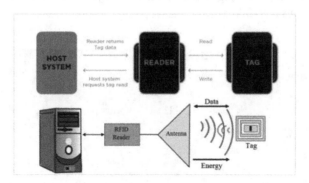

Fig. 1. RFID system bloc diagram

$$r_{max} = \frac{\lambda}{4\pi} \sqrt{\frac{P_{EIRP} G_R}{P_{th}}} \tau \qquad (2)$$

Here, λ is the wavelength, P_{EIRP} is the effective isotropic radiated power transmitted by the reader equivalent to $P_t G_t$ where P_t is the transmitted power and G_t the gain of the transmitting antenna, G_R is the gain of the reception antenna tag, P_{th} is the minimum power threshold to activate the RFID tag chip. While τ is defined as the power transmission coefficient indicated by [5]:

$$\tau = \frac{4R_c R_a}{|Z_a + Z_c|^2} \quad 0 < \tau < 1 \qquad (3)$$

where $Z_c = R_c + jX_c$ is the tag IC impedance and $Z_a = R_a + jX_a$ is the tag antenna impedance.

2.1 UHF RFID Dipole Antenna

The antenna chosen is a dipole antenna, with meandering folds for miniaturization, which is printed on a 25×90 mm^2 epoxy FR4 substrate with $\varepsilon_r = 4.4$ and thickness of h = 1.6 mm. The radiating plane is a copper dipole whose thickness is e = 0.035 mm and the line width is W = 1 mm. After doing a parametric study on the length of all the horizontal bars L and the gap between two horizontal random breaks W1, while fixing the other parameters, we found L = 14.5 mm and W1 = 2.5 mm. In the same way we found the other parameters such as L1 = 32 mm, L2 = 4 mm, L3 = 24 mm, L4 = 3 mm, W1 = 2.5 mm and Wc = 2 mm. The antenna is designed to resonate first in the American frequency band then the insertion of a PIN diode allowed us to cover two bands of frequencies such as the European and American bands by adjusting the state of the diode (ON and OFF) respectively. The technique adopted for the frequency reconfiguration of this antenna is based on the inserting of PIN diode by connecting two stubs at the top of the design Fig. 2. This antenna is suitable for the Monza R6 chip from impinj, which has the following characteristics: (13.5-j126.56) Ohms at 866 MHz and (11.9-j118.4) Ohms at 915 MHz [3].

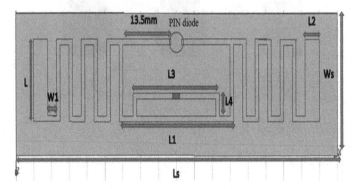

Fig. 2. Reconfigurable UHF RFID dipole antenna

2.2 Antenna Simulations Results Without PIN Diode

With these parameters, the antenna resonates at 916.7 MHz (Fig. 3) with a reflection coefficient of −45.86 dB and an input impedance of 13.02 + j118.88 Ohms which is very close to the conjugate of that of the chip. This means a very good match between the antenna and the chip that will allow the RFID tag to recover the maximum energy that will be sent to it by the reader.

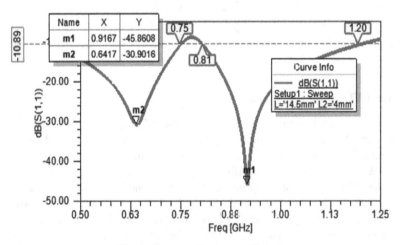

Fig. 3. Power reflection coefficient

Thus, we obtained a gain of 1.9 dB (Fig. 4) and a directivity of 2.1 dB which gives us a radiation efficiency of 90.47%. The antenna covers a wide bandwidth of 390 MHz (810–1200 MHz) for any Power reflection coefficient S11 ≤ −10 dB.

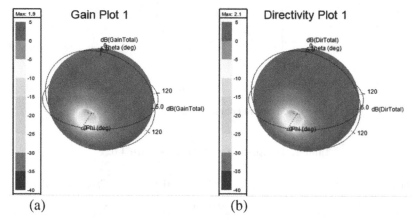

Fig. 4. 3D antenna gain (a) and directivity (b)

3 UHF RFID Antenna Reconfiguration

A reconfigurable antenna is an antenna of which at least one of the characteristics is modifiable after its completion, by application of an electronic component [4]. There are several antenna reconfiguration techniques such as using PIN diode, Varicap diode, MEMS (Micro ElectroMecanical Systems) and optical switches etc.

Our choice on the PIN diode is based on the two frequency bands to be covered, namely the American frequency band and the European one, by playing on the ON and OFF states of the PIN diode.

The diode used is the BAR64-02V whose datasheet is as follows (Fig. 5 and Table 1):

We keep the same antenna parameters by inserting the PIN diode at the level of the two stubs Fig. 2. However, the antenna resonates towards the lowest frequencies (866.7 MHz) in the ON state then in the OFF state towards high frequencies 913.3 MHz around the starting frequency (antenna without diode).

3.1 Results of Simulations of the Reconfigured Antenna

In the ON state, the inductive effect of the diode, which has the role of increasing the electric length of the lines of the folded dipole, allowed us to vary the resonant frequency from 916.7 MHz to 866.7 MHz located in the European band while covering a 380 MHz band at −10 dB with a reflection coefficient of −37.8915 dB Fig. 6. The increase in the reflection coefficient can be explained by losses due to the insertion of an active

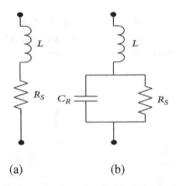

(a)   (b)

Fig. 5. PIN diode equivalent circuit (a) ON state (b) OFF state.

Table 1. Datasheet of the PIN diode BAR64-02V

PIN diode	Resistance (R)	Inductance (L)	Capacitance (C)
ON	2.8 Ω	0.6 pH	-
OFF	3 K Ω	0.6 pH	0.17 pF

component. The insertion of the diode causes a slight variation in the input impedance, which is $14.62 + j119.67$ Ω, which is a little close to the calculated impedance $13.5 + j126.56$ Fig. 7.

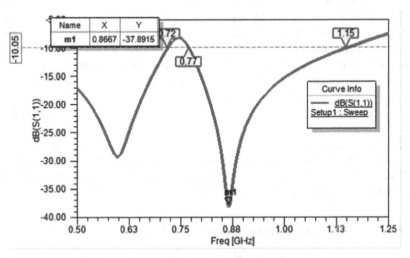

Fig. 6. Reflection coefficient in the ON state

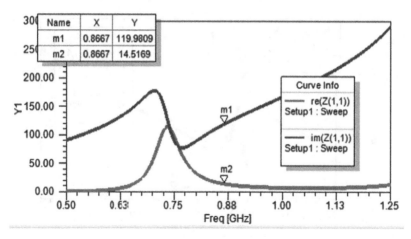

Fig. 7. Antenna impedance in the ON state

It provides a gain of 1.7 dB and a directivity of 2.1 dB which is anonymous with radiation efficiency of 80.95%. Due to the losses caused by the insertion of the diode on the antenna circuit, the antenna loses power which is the cause of the decrease in gain.

Due to the high resistivity of the diode and the capacitive effect which compensates the effect inductive of the diode in the OFF state, the antenna manages to resonate in the American band at 913.3 MHz with a reflection coefficient of −40.53 dB and a band of 400 MHz Fig. 8 as well as a good match of 13.91 + j119.28 Ω Fig. 9.

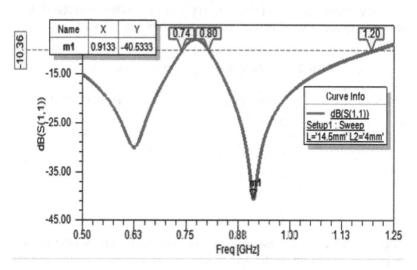

Fig. 8. Reflection coefficient in the OFF state

This antenna in the off state has a gain of 1.8 dB with a directivity of 2.1 dB which means that it is 85.57% efficient. Since the adaptation of the antenna to the state OFF

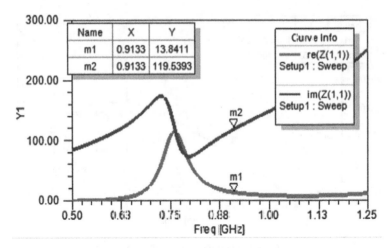

Fig. 9. Antenna impedance in the OFF state

being a little better than that in the ON state with the chip, the antenna gains more power which is the origin of a insignificant increase for the gain increase from 1.7 dB to 1.8 dB.

3.2 Comparison of the Results of the Proposed Antenna with the Results of Some Previous Dual-Band Antennas

The maximum reading range of an RFID tag is determined from the Eq. (2). According to the latest regulatory status of RFID in the EPC Gen2 band (860–960 MHz), the RFID transmission frequency ranges in some countries is shown in Table 2 [7].

Table 2. UHF band frequency and power allocation

Country/Region	Frequency range (MHz)	Power (W ERP)
Europe	865.6–867.6 915–921	4
USA	902–928	4
China	920.5–924.5	2
Japan	916.7–920.9	4

In Europe, the European Telecommunications Standards Institute (ETSI) is responsible for propose regulations in the field of telecommunications. The allowed value for the EIRP in Europe is 3.3 W while in America the EIRP is 4 W. Based on these ETSI standards, we calculated the RFID tag read range at each state. The sensitivity of the chip is −20 dBm [3]. The results are recorded in Table 3.

The comparison between the performance of our work and certain work has been made. The results are reported in the following table.

Table 3. Comparison table of the results of the proposed antenna with the results of some previous dual-band antennas

S. nos.	Substrate	Reconfiguration	Tag volume (mm^3)	Gain (dB/dBi) (865–869 MHz)	Gain (dB/dBi) (902–928 MHz)	Read Range (m) (865–869 MHz)	Read Range (m) (902–928 MHz)
[12]	FR4 époxy	NO	60 × 20 × 1.6	−0 dB	−0.17 dB	5.5	5.8
[13]	FR4 époxy	NO	139 × 32 × 1.6	0.94 dB	0.94 dB	17.4	16.8
This work	FR4 époxy	YES	90 × 25 × 1.6	1.7 dB	1.8 dB	17.7	20.03

These results show that the proposed antenna, with a long read range and its ability to cover two frequency bands, is a good candidate for UHF RFID applications.

4 Conclusion

In this paper, we have proposed a frequency reconfigurable UHF RFID tag antenna capable of operating in two band regions i.e. European band and American band. This antenna printed on an epoxy FR4 substrate is small in size and is a broadband antenna. It can be used in many areas of RFID applications with long range. After simulations, we obtained a gain of 1.9 dB, a directivity of 2 dB, a radiation efficiency of 90.47% as well as a maximum reading range of 20.43 m. The insertion of the PIN diode in the antenna design resulted in low losses on the impedance adaptation as well as for the radiation. Due to a time constraint, the realization of this antenna was not made. In our next publications, we plan to carry out the construction and measurements of the proposed antenna. The reduction in size as well as the use of a ground plane as a reflector would be a major advantage, which will allow us to take measurements in more complex environments.

References

1. Susini, J.-F., Chabanne, H., Urien, P.: RFID and the Internet of Things. Hermes Science Publications, New York (2010)
2. Official Journal of the Republic of Tunisia, vol. 59, p. 2234, 22 July 2008
3. Impinj: Monza r6 datasheet, Monza R6 UHF RFID tag chip (2014)
4. Loizeau, S.: Design and optimization of multifunctional reconfigurable antennas and ultra wide bands. Doctoral thesis, Université Paris-Sud XI, November 2009
5. Ziai, M.A., Batchelor, J.C.: Temporary on-skin passive UHF RFID transfer tag. IEEE Trans. Antennas Propag. **59**(10), 3565–3571 (2011)
6. Casula, G.A., Montisci, G., Mazzarella, G.: A wideband pet inkjet printed antenna for UHF RFID. IEEE Antennas Wirel. Propag. Lett. **12**, 1400–1403 (2013)

7. TGL of Business: Regulatory status for using RFID in the EPC Gen2 (860 to 960 MHz) band of the UFH spectrum. The Global Language of Business (2016)

8. Rao, K.S., Nikitin, P.V., Lam, S.F.: Impedance matching concepts in rfid transponder design. In: Fourth IEEE Workshop on Automatic Identification Advanced Technologies, pp. 39–42. IEEE (2005)

9. Alarcon, J., Deleruyelle, T., Pannier, P., Egels, M.: A new spiral antenna for passive UHF RFID tag on different substrates. In: Proceedings of the Fourth European Conference on Antennas and Propagation (EuCAP), pp. 1–4. IEEE (2010)

10. Gaetano, M.: The art of UHF RFID antenna design: impedance matching and size-reduction technique. IEEE Antennas Propag. Mag. **50**, 66–79 (2009)

11. Riaz, M., Rymar, G., Ghavami, M., Dudley, S.: A Novel Design of UHF RFID Passive Tag Antenna Targeting Smart Cards Limited Area, January 2018. https://doi.org/10.1109/ICCE.2018.8326224

12. Bansal, A., Sharma, S., Khanna, R.: Compact meandered folded-dipole RFID tag antenna for dual band operation in UHF range. Wireless Pers. Commun. **114**(4), 3321–3336 (2020). https://doi.org/10.1007/s11277-020-07530-9

13. Bouazza, H., Lazaro, A., Bouya, M., Hadjoudja, A.: A Planar Dual-Band UHF RFID Tag for Metallic Items. Radioengineering **29**(3), 504-511 (2020). https://doi.org/10.13164/re.2020.0504

Leveraging Cloud Inter-zone Architecture for Response Time Reduction

Birane Koundoul[1][(✉)], Youssou Kasse[1], Fatoumata Balde[1], and Bamba Gueye[2]

[1] University of Bambey, Bambey, Senegal
{birane.koundoul,youssou.kasse,fatoumata.balde}@uadb.edu.sn
[2] University of Dakar, Dakar, Senegal
bamba.gueye@ucad.edu.sn

Abstract. Technology has undergone a rapid evolution in recent years through cluster, grids, cloud and IoT. The latter is leading to the proliferation of data across various domains such as transport, health, environment. In the process, the number of connected devices continues to grow, generating an extraordinary amount of data that the traditional cloud is struggling to manage. In such a situation, the problems encountered revolve around high latency, a decrease in the level of quality of service, high bandwidth, enormous energy consumption. This situation justifies the birth of Fog Computing whose role is not only to collect data from connected objects (phones, vehicles, tablets) in points of presence placed as close as possible to the users. This leads to a reduction in response time. Based on architecture models with Fog nodes in a zone, we propose a new architecture with interconnected zones. It allows us to distribute requests between zones to reduce access to the cloud to reduce latency. Our solution is to interconnect zones in a double ring mode with a set of Fog nodes in each zone. Communication between Fog nodes using the gossip protocol and the distributed hash table for inter-zone communication. We will also propose an algorithm to favour access to Fog Computing over the Cloud. We will detail in the following sections.

Keywords: Fog computing · IoT · Data access · Distributed hash table · Performance evaluation · RdP · Graph theory

1 Introduction

In recent years, the cloud, which was seen as a key infrastructure for on-demand user services encompassing several domains such as commerce, applications, has been challenged by connected objects. These connected objects have undergone rapid evolution and are producing huge amounts of data. In [1,2] the authors estimated that the number of connected devices could reach 50 billion by 2020. This becomes a challenge for the cloud to meet the demands with minimal time. By 2011, smartphone traffic had far surpassed PC traffic. In the US, [2] show that 80% of the population uses smartphones. This confirms that the average number of connected devices per person will reach 6.58 Cisco reported in 2020.

© ICST Institute for Computer Sciences, Social Informatics and Telecommunications Engineering 2021
Published by Springer Nature Switzerland AG 2021. All Rights Reserved
Y. Faye et al. (Eds.): CNRIA 2021, LNICST 400, pp. 87–97, 2021.
https://doi.org/10.1007/978-3-030-90556-9_8

With the Internet of Things (IoT), there is data where real-time responses are expected. Therefore, query processing at the cloud level will not be attractive. In 2012, Cisco proposed a technology called Fog Computing [3,4]. This technology was not born to replace the cloud but to complement it. Because the cloud is far from the users, Fog Computing tries to address the issues of latency, quality of service, mobility.

Fog Computing allows the collection of data from connected objects (phones, vehicles, tablets) in points of presence placed as close as possible to the users and finally send responses at a low time. Fog Computing can be defined as an infrastructure for storing and processing data from connected objects. Fog nodes are heterogeneous in terms of processing performance, storage capacity and latency of access to objects and users. In [5], Kuljeet Kaur et al. argue that the integration of cloud computing and IoT is necessary not only to process the stored data but also to reduce latency. The cloud can process, store a large volume of data but it is somewhat remote from the users. This is why Fog Computing (From Core to Edge) has recently emerged to enable seamless convergence between cloud and mobile for real-time content, data delivery and processing [3,6]. Fog computing is a distributed architecture where data collection points are placed at the edge of users. Nodes are placed in different locations depending on their storage capacity. Those with limited resources are placed closer to the users, while the others (large resources) are placed at the core of the network. There are several research works that focus on latency minimisation among which we can mention: [6,7], on latency reduction based on data placement at Fog nodes. Solutions have been proposed in terms of exact algorithms, heuristics and meta-heuristics. To solve this problem, we propose an architecture with interconnected zones in double ring mode. Our approach differs from existing methods in two important ways. First, it easily reduces the access to the cloud, and second, it also facilitates the communication of the Fog nodes.

The rest of the paper is structured as follows: in Sect. 2, we will review related work. Our architecture is presented in Sect. 3 with a comparison of some existing architectures. In Sect. 4, we define an algorithm to privilege the access to Fog Computing over the cloud. Section 5 presents the experimental results to reduce the access to the cloud. Finally, in Sect. 6, we end the paper with a conclusion and future works.

2   Related Work

Today's networks are complex and contain a large number of heterogeneous components. In fact, the use of Fog Computing can drastically reduce overall network latency [8]. In Fog Computing, the processing and storage components, called Fog nodes, are heterogeneous in terms of performance and storage capacity. In [9], the authors show that Cloud and Fog Computing provide on-demand services but neither of them can guarantee the quality of service of IoT based on delay sensitive applications alone.

Abedi et al. used a delay-sensitive application as a case study, which aims to monitor the health status of people in maritime environments. In this paper, Abedi et al. discuss an artificial intelligence-based task distribution algorithm (AITDA) using a broker between users and servers. This broker is responsible for receiving computing tasks from users and assigning them to servers. The limitation with this architecture is that the broker is not able to distinguish the best resource between the Fog and the Cloud. Furthermore, if the number of tasks available in the broker increases (the number of data arriving), the allocation of tasks to servers will take time.

In [10], Mostafa et al. proposed an algorithm for automated selection and allocation of Fog Computing resources (FResS). In this paper, they also proposed a model for predicting the execution time of a task. The FResS technique stores historical user and device (IoT) data by creating execution logs that can be used for future tasks. In this model, a new layer is placed between the Fog layer and the connected objects which will increase the number of hops.

In [11], Vasileios et al. proposed two architectures to reduce latency. In their first model, the architecture consists of a set of interconnected Fog nodes to facilitate data exchange. They improved their architecture by proposing zones with a set of Fog nodes to approve their storage powers. The limitations with his model are that there is no interconnection of the zones. This avoids direct zone communication. As a result, a request that is not processed in the zone will be redirected to the cloud. This implies a high latency.

According to [12] four graph models are considered the most widely used: the regular graph model, the Erdös Renyi (ER) network model, the Barabasi-Albert (BA) model and the Watts-Strogatz (WS) model. The Erdös Renyi model is a graph that is generated by a random process. In this model, two variants exist and are closely related to the random graph model. ER is a simple yet very rich model, which allows a large number of results to be obtained on large graphs. The number of edges at the vertices can vary. For the Barabasi-Albert model, it is a model that generates a graph with property. This means that some may have more neighbours than others. This model incorporates two concepts, namely growth and preferential attachment. The latter does not exist in the ER model. Growth means that the number of connected nodes increases over time; preferential attachment means that the more connected a node is, the more likely it is to receive new links (edges). It is a network model of scale invariants identifiable by the degree distribution of their nodes (the number of neighbours of a node) [13]. For the Watts-Strogatz model, it is a random graph generation model producing graphs with the small world property. That is, each node is a short distance from all other nodes in terms of hops but the node is not connected to all other nodes [13]. The Watts and Strogatz model is able to address both limitations encountered in ER. It solves the clustering problem while managing the short path between two nodes.

In Sect. 3, we will discuss our model (architecture) of interconnected areas with a set of Fog nodes using the Barabasi-Albert and Watts-Strogatz graphical models to reduce the distance between Fog nodes.

3   Architecture

In Fig. 1, we have represented our architecture with a set of interconnected zones in double ring mode. The main objective is to reduce access at the cloud level. Recall that Vasileios et al. in [11] proposed an architecture with unconnected zones. The latter are connected directly to the cloud, unlike our architecture where the zones are connected to the cloud but inter-zone communication is noted. Also at the level of each zone, several Fog nodes are placed and interconnected. This makes it possible to manage a large amount of user data in a geographical area. A Fog node in a zone is connected to several other Fog nodes in the same zone. The number of neighbours of the nodes is different because with the use of BA and WS models, the number of links between the nodes increases as they are connected. In addition, the gossip protocol is applied at

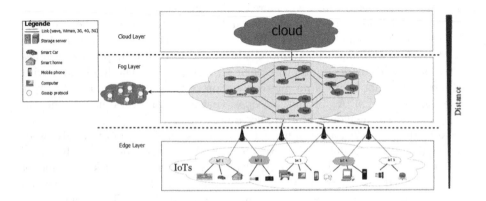

Fig. 1. Interconnection of zones in double ring mode.

each node. This protocol allows the Fog node to know all the objects stored by its neighbours. The neighbours of the Fog nodes communicate the stored information to each other. Each node has a key range which it stores and communicates with its neighbours. This facilitates access to the data. For example, a node A with its neighbours B and C, node A knows the objects stored in nodes B and C. This makes it possible to redirect the request directly to the node that can handle the request.

We have also defined an algorithm for zone switching. This algorithm is detailed in Sect. 4. The objective is to minimise access to the cloud. The redirection of the request in a zone is done by consulting the hash table (DHT). This consultation of the DHT makes it possible to target the zone likely to answer the request. The zones are connected in a double ring. This means that each zone has two neighbours and the advantage is that another link can be used if one of them fails. The principle is the same as for the ring topology, except that an extra ring is added as a backup in case the primary ring fails. At the level of each zone, a controller node is responsible for storing the set of keys (DHTs) of the stored objects in the zone. This controller node communicates with its neighbours (the controller nodes of the neighbouring zones) the stored objects and an update message is sent by the nodes after each new addition or removal of data. Therefore, the controller node will update its hash table.

3.1 Model with Zone Interconnection in Double Ring Mode

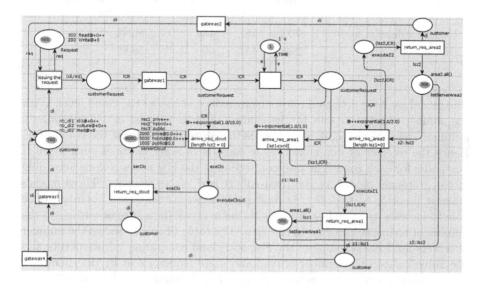

Fig. 2. Representation of the architecture with the petri nets.

Table 1. Number of servers per infrastructure

Infrastructure	Number of servers
Area 1	350
Area 2	300
Cloud	8000

In this article we have adopted the BA and WS graph models. This allows us to find the shortest path between two nodes but also to be able to increase the number of nodes in the network (in a zone) according to the number of connected objects in the zone. We have modelled the system infrastructure as an undirected graph with a set of vertices and edges connecting them. An undirected graph G is given by a section G = (S, A) with S a finite set of vertices and A a set of unordered pairs of vertices $\{s_i, s_j\} \in S^2$. The vertices of the graph are used to model the different Fog nodes existing in the system infrastructure and the edges are used to model the physical links between the different Fog nodes (see Fig. 2).

Table 2. number of requests by type of request

Type of request	Read	Write
Number of requests	300	200

Table 3. Comparative table of the result obtained with the two models

	Area 1	Area 2	Cloud	Lost	Maximum execution time (ms)
Model Vasileios et al. in [11]	200	161	139	0	96.44
Our model	202	191	60	47	56.85

The model we have proposed allows us to reduce access at the cloud level. With the number of requests sent by the connected objects, some of which require a minimum response time, it is preferable to reduce the access at the cloud level. Hence the interest of our model, as it consists of a set of zones with a set of Fog nodes for each zone. The interconnection of the Fog nodes is done according to the graphical models of Barabasi-Albert and Watts-Strogatz. These models make it possible to find the shortest path between two nodes but also to be able to increase the number of nodes in the network (in a site) according to the number of connected objects in the area. The table (Table 1) shows the number of servers used in each zone of the Fog and cloud computing infrastructure and the Table 2 shows the number of read and write requests sent by the client. We have a comparison of the results obtained after simulation in the Table 3.

4   Algorithm to Favour Fog Computing over Cloud

In our algorithm, the client after sending the request, a gateway check is done to determine if the request will be forwarded or returned to the client. The request is lost if it is returned to the client. If not, it will be forwarded and this will allow it to be sent to a Fog node or to the cloud. When transmitting the request, our algorithm favours Fog nodes over the cloud. This reduces the response time of the request. A Fog node is available if it has the resources to satisfy the request. However, each node is located in an area covered by a base station (BS). In addition, at the level of each node, the gossip protocol is applied which allows the nodes to know all the objects stored by its neighbours. Therefore, if the target node cannot process the request, it will forward the request to its neighbour that can process the request. This process remains in the same zone (same base station) because in each zone, there is a node that plays the role of controller. This allows it to know all the objects stored in the zone. After a consultation of its distributed hash table (DHT), this controller node will redirect the request to another zone (a susceptible zone). This change of zone implies the change of base station. At the time of the change, two cases can happen: either the request is lost if the change was not successful or the request is transferred to another zone

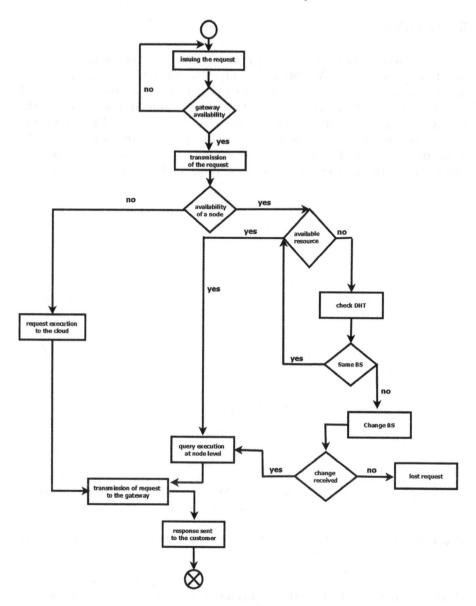

Fig. 3. Algorithm to prioritise access to Fog Computing over the Cloud

(successful change). At the moment the change of zone has been successful, the
same process is applied as the processing of the incoming zone. After processing
the request at a Fog or Cloud node, the response is transferred to the gateway
so that the gateway can forward the response to the client. The advantage of
our algorithm is that it reduces access to the cloud which is far away from the
end users (Fig. 3).

5   Results

After simulation with the CPN tools, we compared the model of Vasileios et al. in [11] with our model. We obtained results that allow us to judge our model as interesting. We placed a limited number of servers in Fog zones (see Table 1) for a number of 500 requests of different types (see Table 2). Figure 4 shows the number of requests executed in the two zones and in the cloud. This shows that there is not a big difference in the number of requests executed between the zones and the cloud. A large number of requests are executed at the cloud level which implies the high response time (see Table 3). Figure 5 shows the results

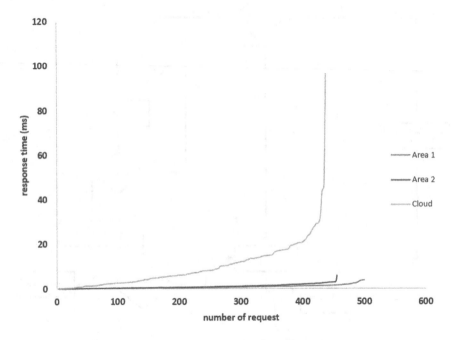

Fig. 4. Response time of 500 requests in an architecture without zone interconnection.

obtained in our model. We notice that compared to the model of Vasileios et al. in [11], a model without zone interconnection (MWHI), we were able to reduce the number of requests to the cloud. This implies the reduction of the response time (see Table 3). In contrast to our model where an interconnection of zones is noted (MWI). As our goal is to reduce the access to the cloud, our algorithm consists in prioritising the requests to the zones for processing. When a gateway is available, we check if a Fog node is ready to process the request, if so, the Fog node processes the request, if not we move to the next Fog node. This check is done as long as the zone contains a Fog node before moving on to another zone. We can say that we have achieved our goal of reducing access to the cloud.

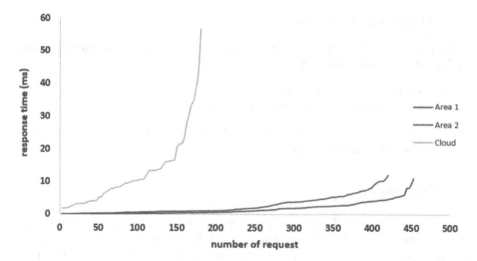

Fig. 5. Response time of 500 requests in an architecture with zone interconnection.

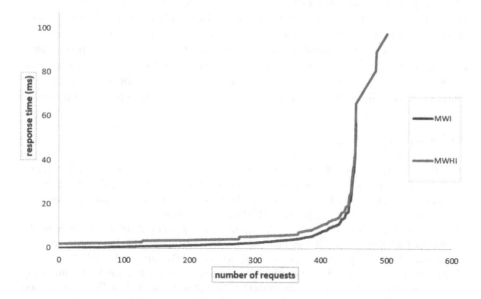

Fig. 6. Performance of both models in terms of response time.

This will allow us to reduce the response time of requests. In Fig. 6, we have the performance of the two models in terms of response time. We see that each model performs well but with some limitations. In terms of response time, our model is better because it allows to reduce the access to the cloud. This is not the case with the MWHI model. In this model, requests will be redirected to the cloud if the zone cannot handle the request.

5.1   Limitations of Our Model

After comparison with the model of Vasileios et al. in [11], we noted some limitations in our model. First, we found:

– lost requests in our model (see Table 3).
– a linear evolution of the response time at the cloud level.

The lost requests are due to the change of zones with a check of the available resources in the neighbouring zones.

6   Conclusion

In this paper, the main idea was to propose an architecture with a set of interconnected zones to solve latency problems. In addition, given the interconnectedness of the zones, an algorithm is proposed to reduce the access to the cloud. This allowed in this simulation to have, a limited number of requests sent to the cloud (60) compared to the model of [11] (139) and a decrease in response time (see Table 3). For the next step, we will try to see how to balance the loads between the Fog nodes and even the zones. In addition we will try to answer the question: how the data placement would be interesting to reduce the response time. Finally, we will try to use an extension of the coloured petri nets which is the hierarchical modular coloured petri nets to take back the zones in the form of modules. This will allow us to reproduce an existing module to create others.

References

1. Alam, F., Mehmood, R., Katib, I., Albogami, N.N., Albeshri, A.: Data fusion and IoT for smart ubiquitous environments: a survey., IEEE Access **5**, 9533–9554 (2017)
2. Luan, T.H., Gao, L., Li, Z., Xiang, Y., Wei, G., Sun, L.: Fog computing: focusing on mobile users at the edge. ArXiv150201815 Cs, mars 2016
3. Bonomi, F., Milito, R., Zhu, J., Addepalli, S.: Fog computing and its role in the internet of things. In: Proceedings of the first edition of the MCC workshop on Mobile cloud computing, Helsinki, Finland, août 2012, pp. 13–16 (2012)
4. Gupta, H., Dastjerdi, A.V., Ghosh, S.K., Buyya, R.: iFogSim: a toolkit for modeling and simulation of resource management techniques in Internet of Things, edge and fog computing environments., ArXiv160602007 Cs, juin 2016
5. Kaur, K., Garg, S., Kaddoum, G., Ahmed, S.H., Jayakody, D.N.K.: 'En-OsCo: energy-aware osmotic computing framework using hyper-heuristics. In: Proceedings of the ACM MobiHoc Workshop on Pervasive Systems in the IoT Era, Catania, Italy, juill. 2019, pp. 19–24 (2019)
6. Stojmenovic, I., Wen, S., Huang, X., Luan, H.: An overview of Fog computing and its security issues. Concurr. Calc. Prat. Exp. **28**(10), 2991–3005 (2016)
7. Taneja, M., Davy, A.: Resource aware placement of IoT application modules in Fog-Cloud Computing Paradigm. In: Conference: 2017 IFIP/IEEE Symposium on Integrated Network and Service Management (IM), pp. 1–7 (2017)

8. Naas, M.: iFogStor: an IoT data placement strategy for fog infrastructure. In: Conference: 2017 IEEE 1st International Conference on Fog and Edge Computing (ICFEC), pp. 1–8 (2017)

9. Abedi, M., Pourkiani, M.: Resource allocation in combined fog-cloud scenarios by using artificial intelligence. In: 2020 Fifth International Conference on Fog and Mobile Edge Computing (FMEC), avr. 2020, pp. 218–222 (2020)

10. Mostafa, N., Ridhawi, I.A., Aloqaily, M.: Fog resource selection using historical executions. In: 2018 Third International Conference on Fog and Mobile Edge Computing (FMEC), avr. 2018, pp. 272–276 (2018)

11. Moysiadis, V., Sarigiannidis, P., Moscholios, I.: Towards distributed data management in fog computing. Wirel. Commun. Mob. Comput. (2018)

12. Khan, B.S., Niazi, M.A.: Modeling and analysis of network dynamics in complex communication networks using social network methods. ArXiv170800186 Cs Math, août 2017

13. Albert, R., Barabási, A.-L.: Statistical mechanics of complex networks. Rev. Mod. Phys. **74**(1), 47–97 (2002). https://doi.org/10.1103/RevModPhys.74.47

IoT and ICT Applications

eFarm-Lab: Edge AI-IoT Framework for Agronomic Labs Experiments

Youssouph Gueye and Maïssa Mbaye[✉] [iD]

LANI (Laboratoire D'Analyse Numérique et Informatique),
Université Gaston Berger de Saint-Louis, Saint-Louis, Senegal
{gueye.youssouph1,maissa.mbaye}@ugb.edu.sn

Abstract. Agronomists deal with challenges to determinate ideal parameters (e.g., soil moisture, temperature, etc.) to grow each variety of plants according to the nature of soil and climate zone. Traditional method consists in having experimental farms in which different conditions are created to discover which environmental and chemical conditions enable maximizing yield for each variety of seed. This process is fastidious and accuracy of results is difficult to evaluate. In this paper, we propose an Edge AI Internet of Things (IoT) framework for agronomic experimentations and will the solution be cost efficient, easy to deploy, low maintenance, and robust, which makes it very appealing in the African context. Our proposal is composed of three segments: experimental farm zone (Lab) where sensors and actuators network are deployed, a set of data collection and processing gateways called Edge AI-IoT Nodes which implements Edge Machine Learning Models, and Cloud and Fog segment that provides a social network and services for agronomic experts. Social network is an interface for agronomic experts that allow them to follow data collected from experimental farms and for cross validation of results around the world. For the purpose of illustration two use cases are presented: plant leaf disease detection using machine learning; and smart automated irrigation with IoT framework.

Keywords: AI-IoT · Edge computing · Smart-agriculture · Smart-irrigation · Machine learning · Experimental farms · Plant Leaf Disease Detection · ICT4D

1 Introduction

Internet of Things (IoT) based Smart-Agriculture is a fast-emerging research and development field with wide range of applications. It consists in using in farming sensors or Unmanned Aerial Vehicles (UAV) to collect data on farm's physical environment (soil moisture, pH., temperature, wind speed, electrical conductivity, etc.) and actuators connected to communication system. The result can be a decision-support systems (such as proper amount of nitrogen, phosphorus, potassium, etc.), optimization system of farming resources (water, fertilizers,

© ICST Institute for Computer Sciences, Social Informatics and Telecommunications Engineering 2021
Published by Springer Nature Switzerland AG 2021. All Rights Reserved
Y. Faye et al. (Eds.): CNRIA 2021, LNICST 400, pp. 101–112, 2021.
https://doi.org/10.1007/978-3-030-90556-9_9

insecticides, etc.) [5], farming monitoring systems (such as detecting plant stress, wheat diseases, pests, and weeds), automated irrigation system.

However, the optimizing tasks and early agronomic research are less addressed by IoT in specific areas in Africa. Indeed, African agronomic researchers and engineers have less opportunities to experiment with a large variety of Farms-Labs environments a large variety of Farms-Labs environments. IoT and AI tools on the edge can be very valuable [13,17]. For instance, an IoT based automated irrigation system wouldn't be efficient without taking into account threshold of dry and moisture that can be supported by plants in the field. Agronomic Engineers might need systems that assist them to monitor their testing farms and provide support in analyzing produced data. This is even more relevant in the African context where there is a lack of agronomists experts and an inefficiency due to outdated and/or out of context data.

Traditional IoT architectures composed of IoT Core network and cloud computing resources are not suitable in the Sub-Saharan Africa area. The main reasons include the followings: firstly these architectures require centralization in a Cloud as well as good network coverage in the experimental fields [14]. Secondly, rural areas in Sub-Saharan Africa suffers from low network coverage and available bandwidth. This makes it very difficult to consider developing centralized architecture. Finally, national agronomic research structures do not have much means to support large scale tests over a long period of time.

In this paper we propose an Edge AI-IoT framework for experimental agriculture that we call eFarm-Lab. Basically, the use of IoT, Edge, and AI in agriculture is not new [7,16]. However, in our knowledge, using a framework for studying agronomic conditions in experimental farms is something new as far as we know. The general principle is that the framework is designed to allow, on the fly, machine learning modelling and deployment of models on Edge Nodes to assist local agronomic researchers in their experimental labs. So the outcome of this proposal targets experimentation farms, not production ones. For instance, to study growth phases of a plant, sensors (cameras, humidity sensors, etc.) can be deployed to monitor the height of the plant and other agronomic parameters, and use machine learning models to better know the needs of the plant.

eFarm-Lab is composed of three segments : Simple IoT sensors and actuators network; a set of Edge-AI-IoT nodes implementing machine learning models for experimentation; and finally Cloud architecture. A social network of agronomic experts as oracles can help labelling data and enhance quality of learning. Social network is an interface for agronomic experts that allow them to follow plants evolution using pictures captured by platform and for cross validation of results by agronomist community.

2   Related Works

There are several works in the field of smart agriculture based on IoT eventually with AI. Topics covers from Smart Irrigation systems [3,9,12], Monitoring and information collection systems [9,11], Crops Protection systems and data analysis [15] and plan disease detection [4]. Main network technologies are WiFi, WiMAX, LR-WPAN, GSM-Based, Bluetooth, LoRa, SigFox, NB-IoT.

Bu et al. [6] use deep Reinforcement learning in IoT network for Smart agriculture. In this work, computations are centralized on the cloud.

Angelopoulos et al. in [2] propose an Edge computing architecture to reduce the traffic between the IoT network and cloud.

Ahmed Imteaj et al. proposed a system that is able to detect the appropriate time to water the field according to the soil moisture and the intensity of light. The system can also monitor irrigation level to prevent accumulation of water around tree roots and send a text message to the farmer in case of lack of water [12]. In [3], authors presents different technologies that can be used in the implementation of an automatic irrigation system for saving water using the Internet of Things. In this article, authors use Zigbee for communication between the sensors and the actuator. Authors of paper [9] designed a basic system based on the Internet and the cloud technologies. LI-FI technology is used to provide communication between the sensors and the data collection server. It is used to collect all the information on the field and to send on the cloud using GPRS or WIMAX as a transmission medium.

[1] has proposed smart farming using automation and IoT technology. The authors have implemented a GPS-based remote-controlled vehicle that will perform several tasks in the field and in the warehouse. Her tasks include scaring birds and animals, detecting soil moisture, spraying fertilizers and pesticides, weeding, detecting soil moisture, and so on.

If we sum up, all these papers about smart agriculture try to enhance agriculture inside production farms. However, before automating irrigation, or detecting plant disease, thresholds must be tested out by agronomist engineers.

Our objective in this paper is to design an Edge AI-IoT framework for experimenting farm conditions for development of varieties of plants in an uncontrolled environment. This point is very relevant for Sub-Saharan Africa since there is not enough agronomist experts to realise this kind of experimentation.

A second aspect of our proposal is it includes a social network of experts in agronomy in order to test different conditions of farming and remotely validated the best ones.

3 Our Proposition

3.1 General Architecture of Proposed Framework

eFarmLab is composed of three segments (Fig. 1): Experimentation farm domain that contains sensors and actuators network, Edge AI-IoT Nodes for small AI training and model deployment and Cloud/Fog for larger machine learning models training and dataset storage.

The experimentation farm area contains a set of plant squares, each corresponding to a specific agronomic experience (seed selection, disease study, plant need, etc.). The different squares can reproduce the same conditions or environment for experimenting and/or monitoring plant evolution (Fig. 2). These area contains plants and network of end nodes (sensors and actuators network). Sensor and actuator network is a set of nodes that embed sensors for collecting

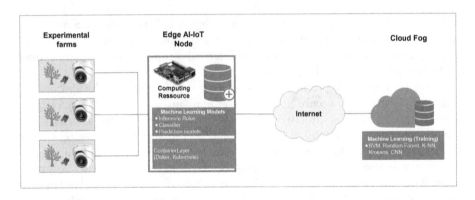

Fig. 1. General architecture for Edge AI-IoT framework

data or modifying environmental conditions via actuator to create different conditions in experimental field (such as starting watering). This network collects data about plant environment parameters like thresholds, soil moisture, plant appearance with camera, and communicate with Edge AI-IoT Nodes.

These sensor-actuator-nodes can manage multiple sensors depending on the need of monitoring. More concretely, if expert wants to explore the stress level of the plant according to the aridity conditions on the maps, it would be possible to use one or more cameras to monitor the general appearance of the plant.

The Edge AI-IoT Nodes (Gateway) act as interface between the sensor and actuator network, and the agronomic experts web based interface through the Internet. These nodes have the role of hosting the intelligence of the network. Intelligence is represented by training lightweight machine learning models but also receiving the deployment of models that come from the fog computing part.

This can be implemented by existing Edge AI platforms (Raspberry Pi, Nvidia Jetson Nano, etc.) with or without a GPU. Tiny Machine Learning models can be trained directly on Edge AI-IoT Node for ROI (Region of Interest) detection (KNN, KMeans, etc.). We will provide an example of directly trained machine learning model on Edge Node.

To allow the deployment of more complex models these nodes host environment containers so that they do not have compatibility issues in running or deployed models. This enables hot deployment and programming of the AI-IoT Edge node.

Finally the Fog/Cloud segment has more computing and storage resources to store larger datasets and train more complex/greedy machine learning algorithms such as CNN. The output models can be deployed on the Edge AI-IoT Nodes.

Agronomists experts use web based interface (social network) to evaluate the result of the machine learning services or enhancing them. With this platform, it could be allowed expert to participate in experimentations by comparing aspects of the plants at different moment. In this way, they can indicate to platform if it is doing well or not. Access devices could be tablets, smartphones and computers.

Fig. 2. Example of experimental farm captured at Saint-Louis/Senegal

The aim of this overall architecture is to automate testing and monitoring for agronomist while giving them possibility to participate on model enhancement.

3.2 Machine Learning Deployment on the Edge IoT

Machine learning tools are deployed in different places in the network. On the Cloud-Computing/fog part where there are more storage and computing resources available, machine learning algorithms are trained on large data sets in order to produce the best models according to what the agronomist expert seeks to study. For example, if the objective of the platform is to detect the presence or absence of disease of a plant from leaves in the Fog part, we will have a dataset of leaves of diseased or healthy plants. These models are created using well-known machine learning algorithms such as SVM, KNN, KMeans, CNN Machine learning models are mainly represented as classifiers, decisions trees, equations, inference rules, etc.

Once the model is validated, it can be deployed to any Edge node it has enough resources. Deployment can be done using containers instead of traditional virtual machines because they are lighter to deploy and consume less computing and storage resources.

This functionality makes it possible to deploy machine models in any equipment, knowing that the context. In the next session we will illustrate use cases of the deployment.

3.3   AI on the Edge for Experimental Fields

Edge KMeans Kernel-Learning for Plant Leaf Disease Detection. Plant
diseases result in an alteration of the plant which modifies or interrupts its
vital functions such as photosynthesis, transpiration, pollination, fertilization,
germination, etc. Manifestations of the disease are usually seen on the leaves,
fruits and stems of the plant. This can have a very big impact on the yield of
the plant. In this use case we consider the context of an agronomist who wants
to study a disease that manifests itself in the leaves automatically.

Edge AI-IoT Node, in this case embed a KMeans Kernel-Learning model to
assist agronomic expert for plant leaf disease detection after a short number of
interactions with the system. The principle of KMeans Kernel Learning consists
in creating KMeans models trained with selected images (Kernel Images). The
clusters resulting from these Kernel Images are called Kernel Clusters and are
then labelled diseased zones or healthy zones [8]. This would help the expert
extracting diseased area even if it is almost visible (Fig. 3).

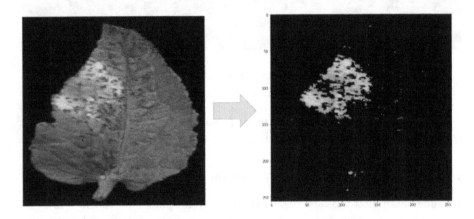

Fig. 3. KMeans Kernel learning clustering

More formally, considering I_{k0} a Kernel Image chosen we have the Eq. (1):

$$KMeans(I_{k0}) = \{\varphi_{k0}, \Omega_{k0}\} \tag{1}$$

Where φ_{k0} is the KMeans Kernel Model clustering model based on the I_{k0}
kernel image and Ω_{k0} is the set of cluster centroids and their labels as a *healthy*
or *disease* regions of the plant leaf. Ω_{k0} is defined by Eq. (2):

$$\Omega_{k0} = \{(\omega_{i,k0}, label)/i \in [0-3], label \in \{health, diseased\}\} \tag{2}$$

Where $\omega_{i,k0}$ is the centroid of the cluster number i (related to I_{k0}) and the
label indicates if the cluster formed from this centroid belongs to a *diseased*
region or *healthy*. We make the assumption that by taking the one cluster that

contains most significant disease region we can make decision about the health of the plant leave. So only one cluster is labelled diseased and we always refer to it by $\omega_{2,k0}$. The framework uses Kernel Image I_{k0} which is supposed to have representative features of a diseased plant leaf. This Kernel Image is used to build a KMean Kernel Model φ_{k0} and Kernel Clusters $\omega_{i,k0}$; i $\in$ [0, 3]. Each cluster can be labelled *healthy* or *diseased*. In our context we orient KMeans algorithm so the cluster that contains most of diseased region is always named $\omega_{2,k0}$. KMean Kernel Models are just classifiers based on KMean that have been trained with data $\mathbb{R}^3$ composed by Kernel Image pixels components (Fig. 4). We limited the number of clusters to 4 because we observed that the number of empty clusters increases when $K \geq 4$.

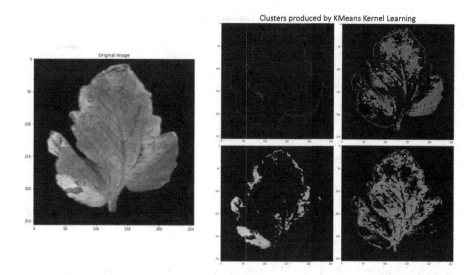

Fig. 4. Segmented plant leaf with four clusters

The KMean algorithm identifies the cluster containing the largest diseased part. Agronomic experts can at this stage tag a few clusters of a few plants to indicate which ones represent a diseased part. The system can present in the social network clusters such as in Fig. 4 so they can retag them if necessary.

As experimentation we use Plant Village DataSet [10] wich is composed of plant leaf images that are segmented. The aim was to design a model for plant leaf disease detection based on Kernel KMeans. We selected 1474 images of diseased plant leaves and 1129 images of healthy plant leaves. For the training/testing split we used 80%/20%. For implementation of Kernel KMeans we used popular Sci-kitLearn, Pandas, openCV and matplotlib.

The results of the test on multiple samples of plant leaf images is presented by the following table.

The model is fast and accurate without much help from experts. The precision is about 95% while accuracy is 93%. It exists machine learning models that are

	Precision	Recall	F1-score	Support
Diseased leaves	0.93	0.95	0.94	1474
Healthy leaves	0.93	0.90	0.91	1129

more accurate and the purpose was not to compete in term of accuracy. However it can be a decision support and monitoring tool for agronomic expert.

3.4 Use Case: Smart Irrigation Experimental Farm

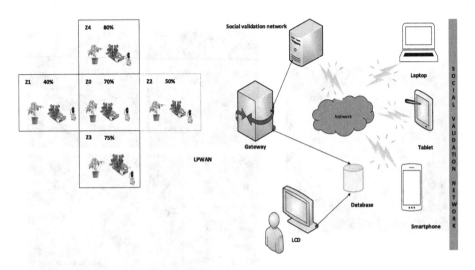

Fig. 5. Architecture for Edge AI-IoT network smart irrigation experimentation

Each plant has its ideal environmental conditions for instance for tomatoes soil moisture should be between 60% and 80%. To discover this kind of information, agronomic tests are done in specialized experimentation farms where different environment conditions can reproduced. Challenging task is to reproduce results for a large number of plant varieties in uncontrolled outdoor environment. To address this problem, as a second use case we propose that experimentation fields are divided into five numbered zones (Fig. 5): Z_0 to Z_4. In each zone, we have a sensors and actuators network that help to learn thresholds for ideal conditions. All zones are connected to one gateway. With this layout, system should learn four thresholds (Fig. 5):

- Z_0: this is the reference zone of our field with a soil moisture threshold that can ensure a good development of the plant. We will therefore compare this zone (Z_0) with the other zones (Z_1, Z_2, Z_3, Z_4) to see which zones have a humidity that favors or alters the appearance of the plant;

- Z_1: the minimum threshold that negatively affects the appearance of the plant with a humidity of 40% compared to Z_0;
- Z_2: in Z_2 the minimum moisture threshold that retains the same appearance of the plant as that found in the reference zone;
- Z_3: the limit threshold which makes it look better than that of the reference plant;
- Z_4: the optimal threshold which gives a very good appearance and a better qualitative transformation of the plant compared to Z_0;

Figure 6 below illustrates the result that our algorithm should provide after the experimental phase of studying our plant. The experimentation is done almost remotely.

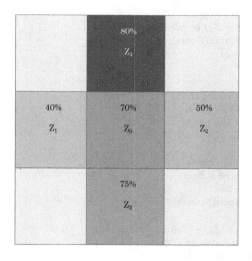

Fig. 6. Example of expected result in the case of tomato

In this use case, model is an algorithm executed by Edge node and that have double inputs: one from sensor network, other from social network of agronomic experts. The algorithm collects data from sensor network and makes decisions according to feedback events from experts in social network part. Indeed, when sensor network does an action (start watering for instance), after a while, it needs to get feedback from agronomic experts which are considered as oracles to tell if this action has positives effect or not. Algorithm 1 executes a main loop.

Algorithm 1: Gateway's Automated Irrigation System Control Algorithm

---

Data:
feedback : feedback of the social network about plant
$Zone_i$: Concerned Zone inside expermation zone
socialNetworkServerAddress : Address of Expert FrontEnd Server
h : local soil humidity/moisture
Result: Thresholds in Z_1, Z_2, Z_3, Z_4
initialization;
while *true* **do**
    picture ← getPicture($Zone_i$);
    send(picture, $Zone_i$, socialNetworkServerAddress);
    feedback ← getFeedback($Zone_i$, socialNetworkServerAddress);
    if *feedback* == *state*$_1$ **then**
        if *($Zone_i$ == Z_1) or ($Zone_i$ == Z_2)* **then**
            | stopWatering($Zone_i$);
        end
        if *($Zone_i$ == Z_3) or ($Zone_i$ == Z_4)* **then**
            startWatering($Zone_i$);
            wait(1 hour);
        end
    end
    if *feedback* == *state*$_2$ **then**
        if *$Zone_i$ == Z_1* **then**
            $h ← h - 5\%$;
            sendToSocialNework($Zone_i, h$);
        end
        if *$Zone_i$ == Z_2* **then**
            stopWatering($Zone_i$);
            $h ← h - 5\%$;
            sendToSocialNework($Zone_i, h$);
        end
        if *($Zone_i$ == Z_3)or($Zone_i$ == Z_4)* **then**
            sendToSocialNework($Zone_i$, FAILURE);
            stopWatering($Zone_i$);
        end
    end
    if *feedback* == *state*$_3$ **then**
        if *$Zone_i$ == Z_3* **then**
            | sendToSocialNework($Zone_i$, h);
        end
        if *$Zone_i$ == Z_4* **then**
            stopWatering($Zone_i$);
            wait(1 hour);
        end
    end
    if *feedback* == *state*$_4$ **then**
        $h ← h + 5\%$;
        sendToSocialNework($Zone_i$, h);
    end
end

First step is the IoT node takes a picture of plant and send it to social network of experts. Agronomic experts give a feedback that is actually of an appreciation of the action of gateway regarding to plant development. The possible feedback events are:

- $State_1$- plant with the same appearance as the reference Z_0
- $State_2$- plant in a state of degradation in comparison to Z_0
- $State_3$- better appearance of the plant in comparison to Z_0
- $State_4$- Substantial improvement in the appearance of the plant in comparison to Z_0.

As an example, when gateway sends a picture of plants in all zones to experts' social network. Experts answer with feedbacks that are represented by Z_i and notifies to the gateway that all plants have the same aspect in all zones ($state_1$). This is to say that the plants have the same appearance in comparison to Z_0 and thus watering is stopped in zones Z_1 and Z_2, but continues Z_3 and Z_4.

4   Conclusion and Future Works

In this paper we proposed an Edge AI-IoT framework for experimental agriculture that we call eFarm-Lab. The general principle is that the framework is designed to allow, on the fly machine learning model training and deployment of models on Edge nodes to asssist agronomic experts in their experimental labs. eFarm-Lab is composed of three kind of nodes: IoT (sensors and actuators) network; a set of Edge-AI-IoT nodes implementing machine learning models; and finally Cloud architecture. A social network of agronomic experts as oracles can help labelling data and enhance quality of learning. We exhibited two use cases where this framework can be deployed for agronomic experimentation. The first is related to plant leaf disease detection and we implemented it to show the proof of concept. And finally the smart irrigation use case to illustrate how the social network of expert can be used to enhance remote testing.

The next step is to deploy a real testbed to see the behavior of deploying an Edge node with controllers encapsulating machine learning models.

References

1. Amandeep et al.: Smart farming using IOT. In: 2017 8th IEEE Annual Information Technology, Electronics and Mobile Communication Conference (IEMCON), Vancouver, BC, pp. 278–280 (2017). https://doi.org/10.1109/IEMCON.2017.8117219
2. Angelopoulos, C.M., Filios, G., Nikoletseas, S., Raptis, T.P.: Keeping data at the edge of smart irrigation networks: a case study in strawberry greenhouses. Comput. Netw., 107039 (2019). https://doi.org/10.1016/j.comnet.2019.107039
3. Anjana, S., Sahana, M.N., Ankith, S., Natarajan, K., Shobha, K.R.: An IoT based 6LoWPAN enabled experiment for water management. In: IEEE ANTS 2015 1570192963, Bangalore, India, pp. 1–6 (2015)

4. Ashok, S., Kishore, G., Rajesh, V., Suchitra, S., Sophia, S.G.G., Pavithra, B.: Tomato Leaf disease detection using deep learning techniques. In: 2020 5th IEEE, International Conference on Communication and Electronics Systems (ICCES), Coimbatore, India, pp. 979–983 (2020)
5. Boursianis, A.D., Papadopoulou, M.S., et al.: Internet of Things (IoT) and agricultural unmanned aerial vehicles (UAVs) in smart farming: a comprehensive review. Internet Things, 100187 (2020). https://doi.org/10.1016/j.iot.2020.100187
6. Bu, F., Wang, X.: A smart agriculture IoT system based on deep reinforcement learning. Futur. Gener. Comput. Syst. (2019). https://doi.org/10.1016/j.future.2019.04.041
7. Calo, S.B., Touna, M., Verma, D.C., Cullen, A.: Edge computing architecture for applying AI to IoT. In: 2017 IEEE International Conference on Big Data (Big Data) (2017). https://doi.org/10.1109/bigdata.2017.8258272
8. Gueye, Y., Mbaye, M., et al.: KMeans Kernel-learning based AI-IoT framework for plant leaf disease detection. In: Hacid, H. (ed.) ICSOC 2020. LNCS, vol. 12632, pp. 549–563. Springer, Cham (2021). https://doi.org/10.1007/978-3-030-76352-7_49
9. Mekala, M.S., Viswanathan, P.: A novel technology for smart agriculture based on IoT with cloud computing. In: 2017 International Conference on I-SMAC (IoT in Social, Mobile, Analytics and Cloud) (I-SMAC), Palladam, pp. 75–82 (2017). https://doi.org/10.1109/I-SMAC.2017.8058280
10. Pandian, J.A., Geetharamani, G.: Data for: identification of plant leaf diseases using a 9-layer deep convolutional neural network. Mendeley Data, V1 (2019). https://doi.org/10.17632/tywbtsjrjv.1
11. Prathibha, SR., Hongal, A.: IoT based monitoring system in smart agriculture. In: 2017 International Conference on Recent Advances in Electronics and Communication Technology, pp. 81–84 (2017)
12. Ahmed, I.T., Rahman, M.K.: IoT based autonomous percipient irrigation system using Raspberry Pi. In: 19th International Conference on Computer and Information Technology, North South University, Dhaka, Bangladesh, 18–20 December 2016, pp. 563–568 (2016)
13. Ran, X., Chen, H., Zhu, X., Liu, Z., Chen, J.: DeepDecision: a mobile deep learning framework for edge video analytics. In: 2018 IEEE Conference on Computer Communications (INFOCOM 2018), pp. 1421–1429 (2018)
14. Roopaei, M., Rad, P., Choo, K.-K.R.: Cloud of things in smart agriculture: intelligent irrigation monitoring by thermal imaging. IEEE Cloud Comput. 4(1), 10–15 (2017). https://doi.org/10.1109/mcc.2017.5
15. Sarjerao, R.K.: a low cost smart irrigation system using MQTT protocol. In: 2017 IEEE Region 10 Symposium (TENSYMP) (2017)
16. Shi, W., Cao, J., Zhang, Q., Li, Y., Xu, L.: Edge computing: vision and challenges. IEEE Internet Things J. 3(5), 637–646 (2016). https://doi.org/10.1109/jiot.2016.2579198
17. Huang, Y., Ma, X., Fan, X., et al.: When deep learning meets edge computing. In: IEEE 25th International Conference on Network Protocols (ICNP 2017), pp. 1–2 (2017)

Enhancing Farmers Productivity Through IoT and Machine Learning: A State-of-the-Art Review of Recent Trends in Africa

Chimango Nyasulu[1]([✉])(iD), Awa Diattara[1](iD), Assitan Traore[2],
and Cheikh Ba[1](iD)

[1] LANI (Laboratoire d'Analyse Numérique et Informatique),
University of Gaston Berger, Dakar, Senegal
{nyasulu.chimango,awa.diattara,cheikh2.ba}@ugb.edu.sn
[2] Grenoble, Grenoble, France
assitan.traore@free.fr

Abstract. Agriculture is considered as the main source of food, employment and economic development in most African countries and beyond. In agricultural production, increasing quality and quantity of yield while reducing operating costs is key. To safeguard sustainability of the agricultural sector in Africa and globally, farmers need to overcome different challenges faced and efficiently use the available limited resources. Use of technology has proved to help farmers find solutions for different challenges and make maximum use of the available limited resources. Internet of Things and Machine Learning innovations are benefiting farmers to overcome different challenges and make good use of resources. In this paper, we present a wide-ranging review of recent studies devoted to applications of Internet of Things and Machine Learning in agricultural production in Africa. The studies reviewed focus on precision farming, animal and environmental condition monitoring, pests and crop disease detection and prediction, weather forecasting and classification, and prediction and estimation of soil properties.

Keywords: Internet of Things · Machine Learning · Innovations · Agriculture · Africa

1 Introduction

Crop and livestock farming (hereafter agriculture) is the major food contributor and key to Africa's prosperity and development. For instance, in Sub-Saharan Africa (SSA), agriculture accounts for over 30% of total Gross Domestic Product (GDP) and over 50% of export earnings. Gross Domestic Product is a standard monetary measure of the value of all the final goods and services produced in a

© ICST Institute for Computer Sciences, Social Informatics and Telecommunications Engineering 2021
Published by Springer Nature Switzerland AG 2021. All Rights Reserved
Y. Faye et al. (Eds.): CNRIA 2021, LNICST 400, pp. 113–124, 2021.
https://doi.org/10.1007/978-3-030-90556-9_10

definite time period by a country [1]. Large part of the agricultural production in Africa come from smallholder rainfed production [2,3].

There are several challenges facing the agriculture sector in Africa: climate change, total dependency on rainfall, poor irrigation infrastructure, small land holdings, poor agricultural extension services, limited access to finance, exorbitant prices of farm inputs, and many more [4–7]. Climate change is a major challenge facing the agriculture sector in Africa and countries across the globe. High dependence on rain-fed agriculture renders Africa more prone to climate change's negative consequences. Due to climate change, there is unreliable rainfall, rising temperatures, eruption of pests and diseases, rampant water scarcity, intense floods and prolonged droughts. Generally, these changes have major negative implications on food security in Africa [4,5]. Currently, large number of people continue to suffer from hunger globally and it is more visible in Africa. Realization of maximum crop yield depends on different crop production attributes like soil properties, rainfall, treatment of pests and diseases and fertilizer application. Timely and precise monitoring of these crop production attributes remain critical for informed decision making and realization of maximum crop yield [4,5].

Information and Communication Technologies (ICTs) have been used to solve real life problems in Agriculture. It is noted that ICTs have been used in timely and accurately remote and proximal sensing of crop production attributes. Information and Communication Technologies promote the transformation of agriculture for improved food production greatly [1]. In the subsequent sections a review of applications of technologies (Internet of Things and Machine Learning) in solving real life agricultural challenges in Africa is presented. The fields of Internet of Things (IoT) and Machine Learning (ML) have turn out to be key for solving different agriculture related challenges. The review discourses the application of IoT in precision farming, animal and environmental condition monitoring. Further, the review discourses the application of ML in pests and crop disease detection and prediction, weather forecasting and classification, and prediction and estimation of soil properties.

2 Use of Internet of Things in Agriculture

Internet of Things is widely used in several domains to help humans carry out day to day endeavors. Agriculture is one of the leading domains where IoT is widely used. Generally, IoT refers to the interconnection of different devices (sensors, smart phones, cameras, etc.) using a defined network architecture to collect and transmit data for monitoring, tracking, tracing, process control, etc. [8,9]. In the following sections we present the main areas where IoT has been used in the agriculture domain in Africa.

2.1 Precision Farming

Precision farming is one of the areas in the agriculture sector where IoT is mostly used. As highlighted in [8], precision farming is the "approach to farm management that uses ICTs for monitoring crop and animal status by observing and measuring variables such as soil condition and plant health to ensure that crops and soil receive exactly what they need for improved resource use, productivity, quality, profitability and sustainability of agricultural production".

In Nigeria, a study performed in [11] deployed a system through use of wireless sensors for optimum catfish production. The system assisted farmers to achieve optimum catfish growth by developing a feeding pattern based on the tracked water parameters: temperature, PH, conductivity and turbidity. Results of the study show that lower feed conversion ratio of 0.62 was achieved against 0.67 in the control pond. A system based on Android platform was developed for user interaction. In Malawi, WiPAM system was developed to automate the irrigation process by observing agricultural field soil moisture changes through use of a sensor. Soil moisture readings were then used by the irrigation controller to determine when to irrigate [12]. A study done in Zambia [13], authors proposed a low-cost automatic irrigation control system by reading soil moisture through use of sensors. Table 1 provides a summary of IoT studies in precision farming.

Table 1. Summary of IoT studies in precision farming

Reference	Objectives	Parameters	Sensors	Transmission protocol
[11]	Optimum catfish production in Nigeria	Water temperature, pH, Conductivity, and turbidity	Temperature, turbidity, electrical conductivity, and pH	GSM/GPRS
[12]	Automatic irrigation in Malawi	Soil moisture	Watermark 200SS	ZigBee
[13]	Automatic irrigation in Zambia	Soil moisture	Soil Hygrometer moisture-sensor	GSM/ZigBee

2.2   Animal and Environmental Condition Monitoring

Animal health, crop growth, high quality and quantity yield are meticulously associated with environmental conditions. Timely and accurate environment information is important for planning farming activities. For example, farmers can plan when to sow, apply fertilizer, apply pesticides, harvest and other farming activities to avoid yield losses. Information and Communication Technologies like Internet of Things are now being widely used to monitor different climatic conditions. Internet of Things sensors can monitor and provide precise real-time climatic condition data at reduced cost [14,15].

In Tunisia, [14] presents a wireless sensor network and cloud IoT based Decision Support System which notifies the farmer when Late Blight may first attack potatoes by sending an SMS. In this study a sensor network monitor and report information to the cloud server about temperature and humidity then SIMCAST model assess the risk of Late Blight appearance. In another study done in Nigeria [20], a system was developed to monitor crop field temperature, humidity and soil moisture using sensors. Sensed environment data was sent to the web server for further processing using Internet. Consequently, the system could automatically trigger irrigation. Web and mobile application were developed for monitoring and user interaction.

In Senegal, [21] proposed a system prototype based on low-cost LoRa IoT platform and IoT cloud platform to prevent cattle stealing in Africa. The system notifies the farmer by triggering an alarm if cows are out of the bounds or if the collar is disconnected or damaged through the Internet (IoT cloud) or if a farmer cannot access Internet, information is sent directly to the smartphone or tablet through WiFi or Bluetooth. Table 2 provide a summary of IoT studies for monitoring of environmental condition.

Table 2. Summary of IoT studies for monitoring of environmental condition

Reference	Objectives	Parameters	Sensors	Transmission protocol
[14]	Detect when Late Blight may first attack potatoes in Tunisia	Temperature and humidity	Wasmposte	Internet, ZigBee
[20]	Monitor crop field environmental condition in Nigeria	Temperature, humidity and soil moisture	DHT11, YL 69	Internet
[21]	Prevention of cattle stealing in Senegal	Proximity and disconnected or damaged neck collar	LoRa end-device	Internet, WiFi, Bluetooth

3   Use of Machine Learning in Agriculture

In agriculture and other domains, big data is being collected every day. The word "Big Data" denotes large heterogeneous volumes of data from various sources which can be structured, semi-structured and unstructured [22]. Big data can be rendered not useful unless if it is organized, analysed, and meaningful features are extracted for decision making.

Unfortunately, humans have limited analytical abilities. To safeguard sustainability of the agricultural sector in Africa and globally, farmers need to gain meaningful insights from big data. Use of technology can help farmers extract meaningful insights from big data for informed decision making. Machine Learning is the present technology which is enabling farmers to efficiently analyze and extract meaningful insights from big data [23–25]. Machine Learning is a subdivision of Artificial Intelligence that includes methods, or algorithms, for automatically identifying patterns from data [26]. The upsurge of Machine Learning technologies allows farmers to solve growing number of real-life challenges [26,27]. In the following sections we present the main areas where Machine Learning has been used in the agriculture domain in Africa.

3.1   Pests and Crop Disease Detection and Prediction

In agricultural production, increasing quality and quantity of crop yield while reducing operating costs is key. Crop pests and diseases potentially reduce quality and quantity of crop yield. Currently, the most widely used and adopted approach for crop disease detection in Africa is through necked eye observation by experienced farmers or experts [29,30]. However, relying on expert's necked eyes to detect crop diseases has many drawbacks: less accurate, time consuming, labor intensive and can only be done in limited areas. Use of technology can help smallholder farmers detect, identify and predict crop diseases early without expert intervention. Image processing and ML are the techniques which have been widely used and adopted for automatic crop disease detection, identification and prediction [29,31,32].

A study done by [31] developed a system for detection of banana pests and diseases using Deep Convolutional Neural Network (ResNet50, InceptionV2 and MobileNetV1). Dataset was gathered from the hotspots in Africa and India comprising of about 18,000 field images taken from different parts of the banana plant. Dataset classes were healthy plant, xanthomonas wilt, dried/old leaves, bunchy top disease, black sigatoka, yellow sigatoka, fusarium wilt and corm weevil. Dataset was split into training set (70%), validation set (20%) and testing set (10%) using simple random technique. Results attained demonstrates an accuracy between 70% and 99%.

In Tanzania, work done by [32] presents a Deep Learning technique to detect invasion of tomato leaf miner at early development stages. Convolutional Neural Network architectures: ResNet50, VGG16 and VGG19 were used to train classifiers on 2145 colored health and unhealth tomato images. Dataset was split into 10% for testing and 75:25, 80:20, and 85:25 ratios into training and validation respectively. Results show that VGG16 achieved highest accuracy of 91.9%. A study done by [33] in republic of Benin and DR Congo demonstrates use of Random Forest and Principal Component Analysis in classifying healthy and diseased banana plant images: banana bunchy top disease (BBTD) and xanthomonas wilt of banana (BXW). Dataset was collected from UAV-RGB aerial images (Sentinel 2, PlanetScope and WorldView-2) from DR Congo and republic of Benin banana fields. Results show accuracy of up to 99%. Table 3 provides a summary of pests and crop disease prediction and detection using Machine Learning.

3.2 Weather Forecasting and Classification

Weather forecasting is defined as foretelling the condition of the atmosphere for a specific location using principles of physics, statistics, empirical techniques and technology. It also includes changes on the surface of the earth produced by atmospheric circumstances. Humans have attempted to forecast weather condition informally by using intuition. Weather forecasting is very important for resource management in crop production: crop growth, fertilizer timing and delivery, pest and disease control and crop yield. Correct forecasting of weather is a complex process. Use of Machine Learning simplifies the process of weather forecasting [34].

Table 3. Summary of pests and crop disease prediction and detection

Reference	Objectives	Datasets	Machine Learning models/algorithms	Results
[31]	Automatic detection of banana pests and diseases in Benin	About 18,000 banana images	ResNet50, InceptionV2, MobileNetV1	Accuracy between 70 and 99%
[32]	Automatic detection of *tuta absoluta* in tomatoes in Tanzania	2,145 colored health and unhealth tomato images	ResNet50, VGG16 and VGG19	Accuracy of 91%
[33]	Automatic classification of health and diseased banana plant in Benin and DR Congo	Health and unhealth banana plant images	Random Forest and Principal Component Analysis	Accuracy of 99%

In South West Nigeria, [35] conducted a study to forecast drought events using Artificial Neural Network (ANN) calculated by Standardized Precipitation Index (SPI). Secondary data for Ijebu-Ode rainfall station was obtained from Nigeria Meteorological Agency. Dataset comprised of rainfall from 1975 to 2013, and maximum and minimum temperature, relative humidity, wind speed and sunshine from 1991 to 2012. The dataset was split into 75% and 25% for training and testing the network respectively. Results indicated 24 episodes of severe drought (-1.5 to -1.99), 44 episodes of moderate drought (-1.0 to -1.49) and 16 episodes of extreme drought (SPI below -2). In Ethiopia, a study by [36] was done to classify current and past drought based on Logistic Regression and Primal Estimated sub-Gradient Solver for SVM (Pegasos) using temperature, precipitation and El Niño data from 1953 to 1993. Results show 81.14% accuracy. In South Africa, a study by [37] was conducted to forecast rainfall rates utilizing Backpropagation Neural Network. The model was trained using 108,861 samples of rainfall data collected from 2013 to 2016 in South Africa. Testing and validation data was collected in South Africa from 2017 to 2018. Results of the model show Mean Square Error (MSE) of 6.017 and Correlation Coefficient of 0.8298. Table 4 provides a summary of weather forecasting and classification using Machine Learning.

3.3   Prediction and Estimation of Soil Properties

Soil is a very important resource for successful agricultural production. For example, soil hold water and nutrients which are essential for crop growth. Generally, without soil, it is difficult to grow crops. Performing soil analysis is vital for a farmer to know the properties of the soil. Consequently, a farmer can determine the crop and inputs suitable for the soil to realise maximum yields [38–40]. Accurate estimation and understanding of soil condition can help in soil management. Soil properties play a major role in crop yield variability. Traditional ways of soil analysis can be slow, costly and not suitable for vastly varied soil environments. Machine Learning techniques provide a reliable solution for estimation and prediction of soil

Table 4. Summary of weather prediction and classification

Reference	Objectives	Datasets	Machine Learning models/algorithms	Results
[35]	Predict drought events in South West Nigeria	Meteorological data for rainfall, relative humidity, maximum and minimum temperature, wind speed and sunshine	Artificial Neural Network	24 episodes of severe drought, 44 episodes of moderate drought 16 episodes of extreme drought
[36]	Classify current and past drought in Ethiopia	Meteorological data for temperature, precipitation and El Niño	Logistic Regression and Primal Estimated sub-Gradient Solver for SVM	Accuracy of 81.14%
[37]	Predict rainfall rates in South Africa	Meteorological data for rainfall	Backpropagation Neural Network	MSE of 6.017 and R^2 of 0.8298

properties [41,42]. This section of the review is about Machine Learning application on prediction and estimation of agricultural soil properties.

[43] conducted a study covering Sub-Saharan Africa at 250 m and 0–30 cm spatial and depth respectively for soil macro and micro nutrient content prediction. Ensemble model was developed using Random Forest and Gradient Boosting. The study targeted 15 nutrients: total phosphorus (P), potassium (K), manganese (Mn), organic carbon (C), magnesium (Mg), organic nitrogen (N), calcium (Ca), iron (Fe), boron (B), calcium (Ca), zinc (Zn), sodium (Na), copper (Cu), sulfur (S), extractable phosphorus (P) and aluminium (Al). In addition to remote sensing, soil samples were collected from 59,000 locations. Results show coefficient of determination (R^2) value of between 40 to 85%. Using two-scale ensemble Machine Learning, [44] performed a prediction of African soil nutrients at 30 m spatial resolution using soil samples data. Targeted nutrients are: total carbon, pH, extractable phosphorus (P), organic carbon (C), calcium (Ca), sodium (Na), total nitrogen (N), iron (Fe), Cation Exchange Capacity (eCEC), magnesium (Mg), zinc (Zn), potassium (K), sulfur (S), clay and sand, stone content, silt, bulk density and depth to bedrock (at 0, 20, and 50 cm depth). Varying results from best to poor show accuracy of soil pH at CCC = 0.9 and extractable phosphorus at CCC = 0.654. Overall, at continental scale, land surface temperature showed to be the most important variable for predicting soil chemical variability.

[45] conducted a study In South Africa for land degradation prediction using Random Forest by combining soil data and environmental data: rainfall, soil temperature, soil moisture, evapotranspiration, elevation, slope, aspect and albedo. Notable overall best results of the model are R^2 of 0.86, Root Mean Squared Error (RMSE) of 7.72% and Relative Root Mean Squared Error (RelRMSE) of 12.94%. Table 5 provides a summary of prediction and estimation of soil properties using Machine Learning.

Table 5. Summary of prediction and estimation of soil properties

Reference	Objectives	Datasets	Machine Learning models/algorithms	Results
[43]	Prediction of soil macro and micro nutrient content in Sub-Saharan Africa	Soil samples	Random Forest and Gradient Boosting	R^2 value of between 40 to 85%
[44]	Prediction of soil nutrient content in Africa	Soil samples	Two-scale ensemble	Accuracy levels from best to poor ranged from 0.900 to 0.654
[45]	Prediction of land degradation in South Africa	Soil samples and environmental data	Random Forest	Notable best results are R^2 of 0.86, $RMSE$ of 7.72% and $RelRMSE$ of 12.94%

4   Our Future Work

Weather information and services are more and more being needed by farmers to survive more capably with climate changeability and growing occurrence of extreme weather events like rising temperatures, floods and droughts. Weather forecasting help farmers plan their farming activities in advance and take preventive measures in case of expected adverse weather conditions. Rainfall is the main source of water for agricultural production. Most of the agricultural production in Africa come from rain-fed agriculture. On the other hand, crop diseases largely contribute to loss of yield quality and quantity. To avert loss of quality and quantity of crop yield due to crop diseases, early detection and prediction of crop diseases and its level of severity is important for right and timely intervention.

Therefore, our future work builds on the already existing studies by developing AI based model(s) adapted to the context of long-term rainfall forecasting and crop disease detection and prediction in the northern geographical part of Senegal. Rainfall forecasting will be done by using ML based on several atmospheric parameters like minimum temperature, maximum temperature and dew point temperature. The best performing model will be selected by evaluating performance metrics such as Mean Absolute Error and Root Mean Squared Error. While crop disease detection and prediction will be done by using image processing and ML based on global and local image features. The best performing model will be selected by evaluating performance metrics such as Precision, Recall, F1-score and Accuracy. An Expert Decision Support System will be developed to utilize the output from the models to provide a platform where farmers and researchers will have access to localized rainfall information, expert recommendations and insights for farmer decision support and action. Additionally, farmers and researchers will use this system to detect and predict crop diseases.

5   Conclusion

By using different IoT approaches and ML models/algorithms such as Deep Convolutional Neural Network, Random Forest, Artificial Neural Network, Logistic Regression, Support Vector Machine, Principal Component Analysis etc., researchers have demonstrated how IoT and ML are set to transform the agriculture sector. Internet of Things and Machine Learning innovations have demonstrated to offer enormous potential for enhancing agricultural productivity and development in developing countries by ensuring efficient use of resources. While Internet access and shortage of expertise are notable challenges in emerging and developing countries, falling prices of smartphones and IoT devices are the driving forces for hasty adoption of IoT and ML innovations. Due to the enormous economic potential of IoT and ML, we propose that governments and development partners in developing and emerging countries invest in the development of policies and ways to support IoT and ML innovations.

Acknowledgements. This work is part of the ongoing PhD training supported by the Partnership for skills in Applied Sciences, Engineering and Technology (PASET) - Regional Scholarship and Innovation Fund (RSIF).

References

1. Brandolini, A., Smeeding, T.M.: Income inequality in richer and OECD countries. In: The Oxford Handbook of Economic Inequality, pp. 71–100 (2009)
2. Warnatzsch, E.A., Reay, D.S.: Temperature and precipitation change in Malawi: evaluation of CORDEX-Africa climate simulations for climate change impact assessments and adaptation planning. Sci. Total Environ. **654**, 378–392 (2019). https://doi.org/10.1016/j.scitotenv.2018.11.098
3. Ngwira, A.R., Aune, J.B., Thierfelder, C.: DSSAT modelling of conservation agriculture maize response to climate change in Malawi. Soil Tillage Res. **143**, 85–94 (2014)
4. Aune, J.B., Coulibaly, A., Giller, K.E.: Precision farming for increased land and labour productivity in semi-arid West Africa. A review. Agronomy Sustain. Dev. **37**(3), 16 (2017). https://doi.org/10.1007/s13593-017-0424-z
5. Shah, M., Fischer, G., Van Velthuizen, H.T.: Food security and sustainable agriculture: the challenges of climate change in Sub-Saharan Africa (2009)
6. Ishengoma, F., Athuman, M.: Internet of things to improve agriculture in sub sahara Africa-a case study (2018). https://doi.org/10.31695/ijasre.2018.32739
7. El Baroudy, A.A.: Monitoring land degradation using remote sensing and GIS techniques in an area of the middle Nile Delta. Egypt. Catena **87**(2), 201–208 (2011). https://doi.org/10.1016/j.catena.2011.05.023
8. Patel, K.K., Patel, S.M.: Internet of things-IOT: definition, characteristics, architecture, enabling technologies, application and future challenges. Int. J. Eng. Sci. Comput. **6**(5) (2016). https://doi.org/10.4010/2016.1482
9. Ndubuaku, M., Okereafor, D.: Internet of things for Africa: challenges and opportunities. In: 2015 International Conference on Cyberspace Governance-CYBERABUJA2015, pp. 23–31 (2015)

10. Liakos, K.G., Busato, P., Moshou, D., Pearson, S., Bochtis, D.: Machine learning in agriculture: a review. Sensors **18**(8), 2674 (2018)
11. Bolaji, A.B., Olalekan, A.W., Olanrewaju, O.E.: Precision farming model for optimum catfish production. Am. J. Electr. Electron. Eng. **8**(2), 51–59 (2020). https://doi.org/10.12691/ajeee-8-2-2
12. Mafuta, M., Zennaro, M., Bagula, A., Ault, G., Gombachika, H., Chadza, T.: Successful deployment of a wireless sensor network for precision agriculture in Malawi. Int. J. Distrib. Sens. Networks **9**(5), 150703 (2013). https://doi.org/10.1155/2013/150703
13. Mulenga, R., Kalezhi, J., Musonda, S.K., Silavwe, S.: Applying Internet of Things in monitoring and control of an irrigation system for sustainable agriculture for small-scale farmers in rural communities. In: 2018 IEEE PES/IAS Power Africa, pp. 1–9. IEEE (2018)
14. Tiwari, M.M., Narang, D., Goel, P., Gadhwal, A., Gupta, A., Chawla, A.: Weather monitoring system using IoT and cloud computing. Weather **29**(12s), 2473–2479 (2020)
15. Marković, D., Koprivica, R., Pešović, U., Randić, S.: Application of IoT in monitoring and controlling agricultural production. Acta Agriculturae Serbica **20**(40), 145–153 (2015). https://doi.org/10.5937/AASer1540145M
16. Mondol, J.P., Mahmud, K.R., Kibria, M.G., Al Azad, A.K.: IoT based smart weather monitoring system for poultry farm. In: 2020 2nd International Conference on Advanced Information and Communication Technology (ICAICT), pp. 229–234. IEEE (2020)
17. Khoa, T.A., Man, M.M., Nguyen, T.Y., Nguyen, V., Nam, N.H.: Smart agriculture using IoT multi-sensors: a novel watering management system. J. Sens. Actuator Networks **8**(3), 45 (2019). https://doi.org/10.3390/jsan8030045
18. Islam, M.M.: Weather monitoring system using Internet of Things. Trends Tech. Sci. Res. **3**(3), 65–69 (2019). https://doi.org/10.19080/TTSR.2019.03.55561
19. Foughali, K., Fathallah, K., Frihida, A.: Using cloud IOT for disease prevention in precision agriculture. Procedia Comput. Sci. **130**, 575–582 (2018). https://doi.org/10.1016/j.procs.2018.04.106
20. Ogunti, E.: IoT based crop field monitoring and irrigation automation system. IJISET-Int. J. Innov. Sci. Eng. Technol. **6**(3) (2019)
21. Dieng, O., Diop, B., Thiare, O., Pham, C.: A study on IoT solutions for preventing cattle rustling in African context. In: ICC, pp. 1–153 (2017). https://doi.org/10.1145/3018896.3036396
22. Riahi, Y., Riahi, S.: Big data and big data analytics: concepts, types and technologies. Int. J. Res. Eng. **5**(9), 524–528 (2018). https://doi.org/10.21276/ijre.2018.5.9.5
23. Sharma, R., Kamble, S.S., Gunasekaran, A., Kumar, V., Kumar, A.: A systematic literature review on machine learning applications for sustainable agriculture supply chain performance. Comput. Oper. Res. **119**, 104926 (2020). https://doi.org/10.1016/j.cor.2020.104926
24. Mishra, S., Mishra, D., Santra, G.H.: Applications of machine learning techniques in agricultural crop production: a review paper. Indian J. Sci. Technol. **9**(38), 1–14 (2016). https://doi.org/10.17485/ijst/2016/v9i38/95032
25. Taghizadeh-Mehrjardi, R., Nabiollahi, K., Rasoli, L., Kerry, R., Scholten, T.: Land suitability assessment and agricultural production sustainability using machine learning models. Agronomy **10**(4), 573 (2020)
26. Lehmann, J., Völker, J. (eds.): Perspectives on Ontology Learning, vol. 18. IOS Press, Amsterdam (2014)

27. Meshram, V., Patil, K., Hanchate, D.: Applications of machine learning in agriculture domain: a state-of-art survey. Mach. Learn. (ML) **29**(8), 5319–5343 (2020)
28. Jia, X., et al.: Mapping soil pollution by using drone image recognition and machine learning at an arsenic-contaminated agricultural field. Environ. Pollut. **270**, 116281 (2021). https://doi.org/10.1016/j.envpol.2020.116281
29. Azfar, S., et al.: Monitoring, detection and control techniques of agriculture pests and diseases using wireless sensor network: a review. Int. J. Adv. Comput. Sci. Appl. **9**, 424–433 (2018). https://doi.org/10.14569/IJACSA.2018.091260
30. Nyabako, T., Mvumi, B.M., Stathers, T., Mlambo, S., Mubayiwa, M.: Predicting Prostephanus truncatus (Horn)(Coleoptera: Bostrichidae) populations and associated grain damage in smallholder farmers' maize stores: a machine learning approach. J. Stored Prod. Res. **87**, 101592 (2020). https://doi.org/10.1016/j.jspr.2020.101592
31. Selvaraj, M.G., et al.: AI-powered banana diseases and pest detection. Plant Methods **15**(1), 1–11 (2019). https://doi.org/10.1186/s13007-019-0475-z
32. Mkonyi, L., et al.: Early identification of Tuta absoluta in tomato plants using deep learning. Sci. African **10**, e00590 (2020). https://doi.org/10.1016/j.sciaf.2020.e00590
33. Selvaraj, M.G., et al.: Detection of banana plants and their major diseases through aerial images and machine learning methods: a case study in DR Congo and Republic of Benin. ISPRS J. Photogrammetry Remote Sens. **169**, 110–124 (2020). https://doi.org/10.1016/j.isprsjprs.2020.08.025
34. Frisvold, G.B., Murugesan, A.: Use of weather information for agricultural decision making. Weather Climate Soc. **5**(1), 55–69 (2013). https://doi.org/10.1175/WCAS-D-12-00022.1
35. Oluwatobi, A., Gbenga, O., Oluwafunbi, F.: An artificial intelligence based drought predictions in part of the tropics. J. Urban Environ. Eng. **11**(2), 165–173 (2017). https://doi.org/10.4090/juee.2017.v11n2.165173
36. Richman, M.B., Leslie, L.M., Segele, Z.T.: Classifying drought in Ethiopia using machine learning. Procedia Comput. Sci. **95**, 229–236 (2016). https://doi.org/10.1016/j.procs.2016.09.319
37. Ahuna, M.N., Afullo, T.J., Alonge, A.A.: Rain attenuation prediction using artificial neural network for dynamic rain fade mitigation. SAIEE Africa Res. J. **110**(1), 11–18 (2019)
38. Tittonell, P., Shepherd, K.D., Vanlauwe, B., Giller, K.E.: Unravelling the effects of soil and crop management on maize productivity in smallholder agricultural systems of western Kenya-An application of classification and regression tree analysis. Agric. Ecosyst. Environ. **123**(1–3), 137–150 (2008)
39. Stark, J.C., Porter, G.A.: Potato nutrient management in sustainable cropping systems. Am. J. Potato Res. **82**(4), 329–338 (2005). https://doi.org/10.1007/BF02871963
40. Gruhn, P., Goletti, F., Yudelman, M.: Integrated nutrient management, soil fertility, and sustainable agriculture: current issues and future challenges. Intl Food Policy Res Inst (2000)
41. Gibbons, J.M., et al.: Sustainable nutrient management at field, farm and regional level: soil testing, nutrient budgets and the trade-off between lime application and greenhouse gas emissions. Agric. Ecosyst. Environ. **188**, 48–56 (2014). https://doi.org/10.1016/j.agee.2014.02.016

42. Du Plessis, C., Van Zijl, G., Van Tol, J., Manyevere, A.: Machine learning digital soil mapping to inform gully erosion mitigation measures in the Eastern Cape. South Africa. Geoderma **368**, 114287 (2020). https://doi.org/10.1016/j.geoderma.2020.114287

43. Cooper, M.W., Hengl, T., Shepherd, K., Heuvelink, G.B.: Soil nutrient stocks in sub-saharan Africa: modeling soil nutrients using machine learning. In: AGU Fall Meeting Abstracts, vol. 2017, pp. B53F–2001 (2017)

44. Hengl, T., et al.: African Soil Properties and Nutrients Mapped at 30-m Spatial Resolution using Two-scale Ensemble Machine Learning (2020)

45. Nzuza, P., Ramoelo, A., Odindi, J., Kahinda, J.M., Madonsela, S.: Predicting land degradation using Sentinel-2 and environmental variables in the Lepellane catchment of the Greater Sekhukhune District, South Africa. Physics and Chemistry of the Earth, Parts A/B/C, p. 102931 (2020). https://doi.org/10.1016/j.pce.2020.102931

Mathematical Modeling and Proposal of an Architecture for the Surveillance in the Distance of Similarly Installations

Bala Moussa Biaye[1]([✉]), Khalifa Gaye[1,2], Cherif Ahmed Tidiane Aidara[1], and Serigne Diagne[1,2]

[1] Laboratory of Computer Science and Engineering for Innovation (LI3), Assane Seck University of Ziguinchor, BP 523, Diabir, Senegal
{b.biaye3299,c.aidara3345}@zig.univ.sn, {kgaye, sdiagne}@univ-zig.sn

[2] Laboratory of Sciences of the Engineer, Computer Science and Imaging (Icube – UMR 7357), National Institute of Applied Sciences of Strasbourg, University of Strasbourg, CNRS, Strasbourg, France

Abstract. In our works, we present a remote monitoring system to handle distributed installations. This system makes it possible to solve the analysis part of the problem of remote monitoring of distributed installations. Remote monitoring of distributed installations requires the use of heterogeneous and sometimes complex tools. A mathematical model allows an analysis the structure of distributed installations. This modeling allows us to categorize the installations in order to facilitate the remote monitoring of similar equipments. The purpose of categorization is to know, for each type of equipment, the necessary tools to monitor them. In addition, from an installation, we can deduce the tools needed to monitor equipments same category. Then, we propose an architecture that will be implemented for the remote monitoring of installations distributed on the territory. This architecture highlights many sensors installed in the equipments for the acquisition of measurements datas. The choice to use many sensors is justified by the fact that installation is consist of many equipments of a different nature.

Keywords: Remote monitoring · Mathematical modeling · Sensor · GIS · Distributed installations

1 Introduction

The control of the proper functioning of the installations spread over the territory must be based on a continuous control of their quality in order to quickly find non-compliant situations. This monitoring is provided by a remote monitoring technique. Implementation remote surveillance system requires taking into account many factors (what type of sensors, protocol, database etc., to use). To manage each of the elements of the architectural proposal in Fig. 4, it is necessary take into account the natural state, the heterogeneity of

© ICST Institute for Computer Sciences, Social Informatics and Telecommunications Engineering 2021
Published by Springer Nature Switzerland AG 2021. All Rights Reserved
Y. Faye et al. (Eds.): CNRIA 2021, LNICST 400, pp. 125–130, 2021.
https://doi.org/10.1007/978-3-030-90556-9_11

the distributed installations and their rapid evolution. We want to suggest a mathematical model to analyze the structure of distributed installations. The mathematical modeling of installations is to classify installations by category. This allows us for each type of installation to identify these equipments. Moreover, for each type of equipment, what tools are needed for remote monitoring. Remote monitoring applications have been proposed in the literature without use mathematical modeling [1–3]. Biaye et al. [1] works propose an smart method for real time surveillance of reparted infrastructures and equipments in rural zones. Lei Y. et al. works [2], concern on carry to search on information acquisition and broadcasting based on remote surveillance method of a new motor to energy.

A mathematical model allows us to categorize the installations in order to facilitate the remote monitoring of similar equipments.

In Sect. 2, we present a mathematical modeling of distributed installations. In Sect. 3, we present the similarity classification of the equipments to be monitored. In Sect. 4 of this paper, we present the architecture for remote monitoring of distributed installations. And finally, the paper presents the conclusion and offers some development perspectives of the remote monitoring system.

2   Mathematical Modeling of Installations

Installations networks analysis can be defined as the study of a relational phenomenon. In other terms, it is about manipulating installations and making connections between those installations. For this, we can deduct from an installation tools necessary to monitor equipments same category. The modeling of an installation network is part of this logic in our works. This is one or more methods of describing relationships between installations and equipments. In our works, we use the approach based on the characteristics (model, power, capacity, etc.) of the equipments. This approach is a comparison of two installations (Fig. 1). The part common to these two installations contains equipments with the same characteristics in common. And the disjointed part, contains equipments with different characteristics.

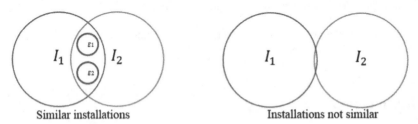

Similar installations   Installations not similar

Fig. 1. Relationship between two distributed installations

Two installations of same type located in differents localities or in same locality may not have same equipments in common. In this context, the need to know the number of equipment in each installation and their characteristics is important to determine the needs in terms of material to monitor them. Each piece of equipment to be monitored must be instrumented with a measurement acquisition system (sensor) for the collection of measurement data. An installation (I) can have one or more equipments (E) (Fig. 2). We identify relations existing between an installation and these equipments. The objective is to group the distributed installations to be monitored by category. In order to deduce from an installation materials necessary for remote monitoring same category installations, we verify similarity of their equipments.

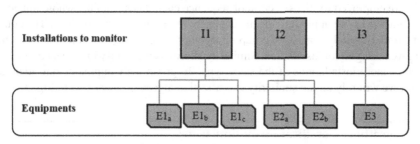

Fig. 2. Tree structure of distributed installations.

In our work, we use the contrast model. In this model, resemblance between entities are determined by a linear combination of measures of the common and distinct attributes for each entity:

E_1 and E_2 are two equipments. $E_1 \cap E_2$ represents characteristics (model, power, capacity, etc.) that E_1 and E_2 have in common, $E_1 - E_2$ represents characteristics that E_1 has but not E_2, in the same way $E_2 - E_1$ represents characteristics that E_2 possess but not E_1. Similarity between E_1 and E_2 is given by formula (1):

$$S(E_1, E_2) = \theta k(E_1 \cap E_2) - \alpha k(E_1 - E_2) - \beta k(E_1 - E_2) \tag{1}$$

Where k is an additive function. θ, α and β are coefficients attributed to the joint and disjoint parts. From this formula, we can identify similarity of two equipments. This allows us to also calculate the total number of similar equipment by the following formula (2):

$$T_E = \sum_{i,j=1}^{N} S(E_i, E_j) \tag{2}$$

Calculating the total number of similar equipment allows us to determine the number of sensors using for categories of installations.

3 Classification by Similarity of the Equipments to Be Remote Monitored

Classification by similarity is a study that is most often used in text mining [4–6]. Kotte, V.K. et al. [4] have proposed a similarity function for feature pattern clustering and high dimensional text document classification. Abualigah, L. M et al. [5] propose three feature selection algorithms with feature weight scheme, dynamic dimension reduction for text document clustering problem. Informative features in each document are selected using feature selection methods.

Ma, A. et al. works [6] propose a new strategy for constructing sequential features from single image in long short-term memory (LSTM) is proposed. Two pixel-wise-based similarity measurements, including pixel-matching (PM) and block-matching (BM), are employed for the selection of sequence candidates from the whole image.

In our works, we apply it to the management of distributed installations. The equipments similarity study consists of grouping together those having the same characteristics for a given category of installation. Similarity allows us to identify identical equipments in two or more installations of the same type. To group the equipments into identical groups, we will rely on the similarity classification algorithm presented in Fig. 3.

Algorithm : Classification by similarity ;
Input :
           N ← Total number of equipments to monitoring ;
Sortie : Similar equipments ;
Begin

For i := 1 to N **do**

$$S(E_1, E_2) = \theta k(E_1 \cap E_2) - \alpha k(E_1 - E_2) - \beta k(E_1 - E_2)$$

   If $S(E_1, E_2) > 0$ then

       E_1 et E_2 are similar ;

       End if

End for
End.

Fig. 3. Equipments similarity calculation algorithm

In order to facilitate the implementation of the remote monitoring application of distributed installations, we have proposed an architecture.

4   Remote Monitoring Architecture of Distributed Installations

The software architecture describes in a symbolic and schematic way differents elements of one or more computer systems, their interrelation and interactions. We propose here an information system architecture for the capitalization of knowledge, the sharing of data for remote monitoring of distributed installations. The functionalities of the application will be implemented according to the architecture shown in Fig. 4. The equipments according to its category is instrumented with acquisition systems capable of acquiring measurement data. The proposed network architecture allows collection of information on the operating state of installations distributed on many remote sites, storage, analysis and real-time treatment of this information with a view to making maintenance decisions of equipments. Moreover, on the other part, to connect the storage database with Arc Gis to allow geolocation of the position of failures equipments.

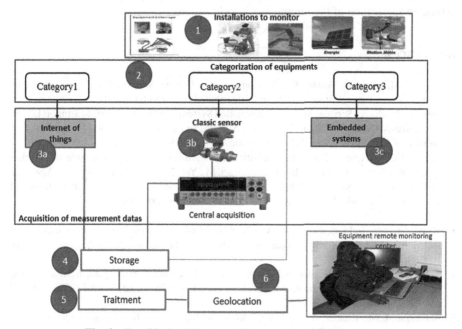

Fig. 4. Simplified architecture of the remote monitoring chain

The architecture proposed in Fig. 4 integrates in a hybrid way the technology of real-time systems, embedded system, connected objects to internet and GIS, for implementation of remote monitoring system. The diversified use of measurement acquisition systems is justified by the fact that we have many equipments of a different nature to remote monitoring. In each equipment must be installed a measurement acquisition system capable of sending data of measurements taken. The real-time system is applied to the treatment of the acquired data. In this system, the respect of the time constraints in the execution of the treatments is as important as the result of the treatment. In fact, after detection of the failure, an alert message is automatically sent to the remote monitoring center. GIS allows us, after detection of the failure, to geolocate installation in order to facilitate rapid access to maintenance teams.

5   Conclusions and Perspectives

This work allowed us to categorize distributed installations and to quantify similarities equipments. This allowed us to identify from one installation the materials necessary for remote monitoring of pairs installations. This identification of materials led to establishment of an architecture. Architecture symbolically and schematically describes different elements of remote monitoring system, their interrelationships and interactions. The architecture proposed in this work integrates in a hybrid way the technology of real-time systems, embedded system, internet of connected objects and GIS, for implementation of the remote monitoring system. Functionalities of the application will be implemented according to this architecture. We envisage as perspectives for this work to finalize application of the remote monitoring system.

References

1. Biaye, B.M., Gaye, K., Aidara, C.A.T., Coulibaly, A., Diagne, S.: Infra-SEN: intelligent information system for real time monitoring of distributed infrastructures and equipments in rural areas. In: Rocha Á., Serrhini M. (eds.) Information Systems and Technologies to Support Learning. EMENA-ISTL 2018. Smart Innovation, Systems and Technologies, vol. 111. pp. 188–193. Springer, Cham (2019). https://doi.org/10.1007/978-3-030-03577-8_22
2. Lei, Y., Li, X.: Research on data acquisition and transmission based on remote monitoring system of new energy vehicles. In: Atiquzzaman, M., Yen, N., Xu, Z. (eds.) Big Data Analytics for Cyber-Physical System in Smart City. BDCPS 2020. Advances in Intelligent Systems and Computing, vol. 1303. Springer, Singapore (2020). https://doi.org/10.1007/978-981-33-4572-0_181.
3. Msayib, Y., Gaydecki, P., Callaghan, M., Dale, N., Ismail, S.: An intelligent remote monitoring system for total knee arthroplasty patients. J. Med. Syst. **41**(6), 1–6 (2017). https://doi.org/10.1007/s10916-017-0735-2
4. Kotte, V.K., Rajavelu, S., Rajsingh, E.B.: A similarity function for feature pattern clustering and high dimensional text document classification. Found. Sci. **25**(4), 1077–1094 (2019). https://doi.org/10.1007/s10699-019-09592-w
5. Abualigah, L.M., Khader, A.T., Al-Betar, M.A., Alomari, O.A.: Text feature selection with a robust weight scheme and dynamic dimension reduction to text document clustering. Expert Syst. Appl. 84, 24–36 (2017). https://doi.org/10.1016/j.eswa.2017.05.002
6. Ma, A., Filippi, A.M., Wang, Z., Yin, Z.: Hyperspectral image classification using similarity measurements-based deep recurrent neural networks. Remote Sens. **11**(2), 194 (2019). https://doi.org/10.3390/rs11020194

Author Index

Printed in the United States
by Baker & Taylor Publisher Services